SCIENCE AND ENGINEERING POLICY SERIES

General Editors Sir Harrie Massey
Sir Frederick Dainton

Science and the media

PETER FARAGO

OXFORD UNIVERSITY PRESS 1976

Oxford University Press, *Ely House, London W.1*

Glasgow Bombay
New York Calcutta
Toronto Madras
Melbourne Karachi
Wellington Lahore
Cape Town Dacca
Salisbury Kuala Lumpur
Ibadan Singapore
Nairobi Hong Kong
Dar es Salaam Tokyo
Lusaka
Addis Ababa

ISBN 0 19 858324 9

© OXFORD UNIVERSITY PRESS 1976

All rights reserved. No part of this publication may be reproduced, stored in a retrieval system, or transmitted, in any form or by any means, electronic, mechanical, photocopying, recording or otherwise, without the prior permission of Oxford University Press

Reproduced and printed by photolithography and bound in Great Britain at Billing & Sons Ltd., Guildford and Worcester

Preface

The importance of the ideas and tangible embodiments of science in making the decisions that are necessary for the conduct of contemporary life is becoming widely recognized. The appreciation of science as an art, or at least as a valid cultural endeavour, belongs to our intellectual traditions. Both these aspects of science imply some degree of general awareness of scientific facts and ideas, an element of common language between the expert and the man on the Clapham bus, between doers and those who only stand and wait. At the meeting-point between the several opposing pressure groups of scientists, politicians, industrialists, and editors, one can observe the slightly contorted figure of the science journalist, performing his many balancing acts. For the purposes of this essay, the term 'science journalist' includes all those who attempt to bring science before a wide public, whatever the medium in which they work, and even if the wide public of one is a small minority of another.

The science journalist is in good company when he observes that the communications of science are in an unsatisfactory state. There are several reasons: some technical, some the inevitable consequences of the growth of science. It must, however, be argued that more reasons than is generally believed arise from bad social and cultural habits. In the last chapter a number of concrete proposals are made for improving the communications of science, principally to those who are not scientists.

One must emphasize that this is not a technical do-it-yourself treatise for crisper copy from cumbersome communicators. Rather it is an attempt to examine some of the factors influencing the science journalist's work and the technical, economic, and conceptual difficulties he has to overcome.

A substantial part of the essay is based on conversations over the past decade with a variety of people working in scientific communications on

Preface

both sides of the Atlantic: scientists, editors of newspapers and journals, professors of journalism, television and radio people. Neither should those go unmentioned who spend their days in the firing line, and the quality of whose contributions is the ultimate statement about scientific communications. There are far too many for individual acknowledgement, and common sense suggests that their views should be reported without naming names. Mistakes of understanding and interpretation are mine alone.

I must, however, offer some particular thanks. The original idea for this book was suggested by Sir Frederick Dainton, F.R.S., and although the choice of implementer might be questioned, the idea itself is proof that the scientific community now takes the problems of communication seriously. My thanks are also due to the staff of *Chemistry in Britain*, past and present, for their heroic if sometimes sisyphean attempts to educate the editor. Last, but by no means least, I should like to thank Miss M. Parker, without whose gargantuan labours at the typewriter this text would literally never seen light of day.

Northwood   P.J.F.
Middlesex
August 1975

Contents

1	News about science	1
2	Is science too difficult?	14
3	Science in the press	29
4	Hard and soft science	42
5	Science and television	52
6	Big science and concerned science	63
7	Two-way communications	72
	References	89
	Index	93

1

News about science

Communications between scientists can be precise and quantitative. Following the development of modern science, a complex of mechanisms has grown up to provide the scientist with the latest information in his field. From journals to computerized information retrieval systems, the scientist is assured of data and explanation which typically will be published within a few months of the announcement of a new result. It has been argued that the exponential increase in the amount of scientific information appearing in print is to the detriment of science; the fact remains that all published information is judged by scientist referees to be of an acceptable quality. The complexities of transmitting scientific information for professional use are for the scientific community to simplify or to accept.

The communication of scientific and technical facts, explanations, and ideas to those who are not scientists is fraught with difficulties of its own. They relate to the nature of the recipients, the nature of the transmission, and the use made of information received. No analysis of scientific communications can succeed without a firm realization that scientists, audiences, and transmission mechanisms are all emphatically heterogeneous. Ideas, traditions, and requirements differ in fundamental ways, and adequate—to say nothing of significant—communications must take these differences into account.

Among the non-scientist recipients of scientific information one can easily distinguish two major groups according to whether they posses a built-in filter or translation agency dealing with the presentation of scientific items. The heads of large industries, government departments, and even some politicians, none of whom are in general scientists, enjoy the services of scientific advisers whose task is primarily to interpret the significance of the state of the art in particular disciplines or fields of science. Whether the interpretation is correct or not, whether it has the desired result, or what the desired result should be, form the discussion of books other than this. Here we shall primarily be concerned with the transmission of information to non-scientists who possess no mechanism for interpreting the significance of scientific advances and who therefore rely on the media for their ideas about science and technology and their significance. There is naturally an overlap between the two populations: established filter mechanisms are designed to interpret science within narrow bands, according to the requirements of a particular office. The company director who might be knowledgeable about the role of enzymes in chemical synthesis might well be totally ignorant of the significance of laser-induced nuclear fusion. More important: knowledge in one area, without a general background of scientific thought and methods, may be totally inadequate to illuminate the significance of science in the much wider context of society.

Whether the non-scientist is well supplied with scientific communications, even disregarding their quality, is a matter of debate. It has been shown, for example, that the amount of science presented on BBC radio has hardly varied over the past fifty years.[1] On the other hand, there are indications that audiences watching television programmes on scientific topics are much larger than expected, and, at least in Britain, have a fairly wide range of presentation to choose from.[64] Science for non-scientists in periodical publications appears to have diminished over the last ten years, although available statistics are not illuminating.[2] It is noticeable that news about science reaches the general public in waves rather than in a steady stream: a topic might be described, discussed, and argued editorially only to disappear suddenly from the pages of the newspapers. During the discussion period the public might be bombarded with conflicting information and advice; the fluoridation of water or the atomic power station debates are memorable examples of the information overkill. What is undoubtedly lacking is a sustained flow of information that would allow aspects of science to become part of the general consciousness.

Why should anybody be concerned about the present state of scientific communications with the general public? In particular, why should many scientists, even if still a minority, take every opportunity to emphasize a need for greater general understanding of science? What advantages can either the scientific community, or society at large, legitimately expect from an improvement in communications?

There are fundamentally two interrelated answers. Science is part of our contemporary culture. The steady attrition of ignorance, the greater awareness that science offers of man and his environment, represent a triumph for the human spirit and endeavour that must be counted among the greatest achievements of mankind. The ability to manipulate concepts, to form patterns, is as much part of the human genius as the ability to express itself through the creation of poetry, painting, and music. It is an enlargement of the human spirit, of human consciousness. It is a part of humanity itself. It is part of our cultural heritage and an indication of our future; to deny the majority the opportunity to take part in this exploration would be to deprive them of the glory, excitement, and spiritual and intellectual development they have every right to as fellow human beings. It is a sad commentary on our present sense of values that, whereas the interpretative arts are deemed worthy of national and local support, the communications of science are left to the care of the individual media.

The advocacy of science as a living part of contemporary culture must not be considered to be an appeal for detailed factual scientific knowledge. The argument about the second law of thermodynamics is here irrelevant; it is to a large extent even irrelevant in science, where many professional scientists exist comfortably without knowledge of it.[3] Our argument does however suggest that the ideas, aspirations, and limitations of science should become matters of general consensus. Particular scientific achievements, their significance, and social consequences can then be considered against such a background.

The second part of an argument for scientific communications must involve the functional use of the science. At the most expedient level there are problems involved in the utilization of science in international, national, and local contexts, which have to be resolved in the political arena. The generation and prosecution of national science policies has been the subject of long debates.[4] The complex involvement of science in the growth of societies has been accepted.[5] But even at the most tangible level, considering the utilization of science by government departments which directly or indirectly enable particular scientific developments to take place, there is no general agreement on the best ways to harness the discoveries of science to the aims perceived by society at large.[6]

Large-scale decisions on the support of science are generally taken by non-scientists. Invariably they are advised by scientists. Yet they must partake of general feelings about science, the results of past understanding or ignorance.

It would be simple if the need for scientific understanding were to be limited to a few decision-makers in high places. It would also be very dangerous, at least for a democratic society. Not only is science a social activity, but its consequences have a deep effect on the future of society; conversely, its direction at any point in time is affected by the real or imagined needs of society. It is necessary therefore to argue the need for a role of scientific awareness that encompasses the whole of society, scientists and non-scientists alike, whatever their functional position. Only in this way can a significant understanding emerge which might take science into account as one of the multitude of factors governing the present and the future.

It is of the essence in science that the future represents undreamed-of opportunities and dangers. All things are possible provided that enough resources are dedicated to their achievement. Such considerations call upon the body of society to make judgement about the actual and potential outcomes of scientific research. Decisions must be made. Such decisions are not fundamentally scientific, although they include science as one of their components. The decisions are made by society on the basis of society's needs; conversely science is not neutral but must be a servant of society.

A corollary is the presence of fears evoked by misunderstandings about the nature of current science and its possible consequences. Such misunderstandings are often blamed on the science journalist, on his love of the scare story, or his attempts to dramatise and vulgarize the solid if unremarkable achievements of scientists. Good examples of this genre have been the wild speculations about genetic engineering and the production of clones. But before the science journalist is allowed to carry the total burden of scientific misapprehension, it is pertinent to ask whether he had been encouraged in his task by the self-important scientist who, blinkered by his own obsessions, implicitly or explicitly attached greater importance to his work than it had merited.

Both these domains of action—decisions about future roles of science and the allaying of fears arising out of current science—demand an intelligent general public that can respond in a rational way to the challenges of scientific and technological research. Its response can take two important forms.

The public should be in a position to question the scientists. It has been said that to obtain answers in science is easy; it is the formulation of questions that is difficult. In the same way, in order to arrive at a proper assessment of science, it is necessary to ask appropriate questions either from scientific advisers or from scientists in general. There must be some understanding of the general thought-processes of science if this is to be done.

Arriving at the correct decision is the logical outcome of appropriate questioning. But in order that the decision process should have any significance, it has to take place in good time, before the inexorable forces of development have gone too far. Decisions have to be made before action has been forced by the sheer momentum of events. A true decision should not be confused with an angry hunt for scapegoats after the deed had been done. In order that correct social action may be taken in matters involving science, an understanding of the effects of science is essential. The more widespread such understanding, the better.

Both the asking of questions and decisions based on assessments about the future consequences of science depend on appreciating the significance of particular scientific discoveries or trends. It is naive to imagine that the scientific community offers an outstanding discovery every day: most scientific endeavour is more mundane. Because there is so much science, because by tradition all scientific results are professionally published as having equal weight, it is all too easy to miss the significance of particular new departures.

It is in this context that the transmission of scientific information is of the greatest importance. Such communication must not be overtly educational, although it must have a didactic purpose. It cannot consist of talking down to an audience defined as unintelligent. It should not even vulgarize—in the worst meaning of the term—by oversimplification and the drawing of tedious parallels between a few facts of science and the everyday experiences of its market.

It should, on the contrary, act as the scientific advisory service to those not privileged enough to possess their own. It should attempt to point out both the glories and the dangers of science. It should insist on the cultural and historical importance of science. It should allay fears but draw attention to future consequences. But above all it should act as a bridge between the thoughts, ideas, and ordering processes of science and those of non-scientific culture, both being facets of our common humanistic traditions. Both science and the arts are attempts at ordering man's universe; they deal with different aspects, but the commonality of principles is greater than is usually appreciated.

This is what scientific communications are essentially concerned with, rather than with attempting to teach the taxonomic differences between insects and spiders or the half-lives of the transuranium elements. Once this challenge is recognized, the social consequences of science and their automatic discussion will follow. So will a constant search for items of true significance that illuminate the future paths of scientific endeavour.

Neither should one disregard the fundamental need for scientific communications in this context to be two-way thoroughfares. It is not sufficient for scientific news to reach the general public; it is also important that the fears, hopes, difficulties, and dreams of the public should be conveyed to the scientist in terms he not only understands but to which he can functionally react. In this way the scientific communicator becomes a bridge between two communities, a catalyst of new attempts at improving the society he inhabits.

The improvement may be material, by focusing attention on some area of neglect, for example industrial health, or no possibilities still to be exploited, as has been the case with new energy sources. But one must emphatically argue the case that improvements in understanding, in appreciation, are more important than material gains. Better understanding tends to establish a new atmosphere, an awareness of science, not only as a materialistic but also as a cultural force whose strength resides not only in discovery, but also in the applicability of its ideas and methodologies to matters far removed from science, but including the social consequences of science itself.

The task of the scientific communicator is therefore to persuade his public to regard science as part of his cultural environment, in the same way that he expects his daily newspapers to carry notices of the theatre, films, and books. His task is to demythologize science, for only then will it be regarded neither with love not hate but with awe and scepticism, as indeed it should.

The idea of news

Scientific communications must be transmitted to the non-scientist in a form largely defined by the requirements of the media. The word for this package is news, although one should bear in mind that one man's news is another's old hat. News is the peg from which virtually all communications must hang, with obvious exceptions such as the rare, broadly based historical feature. The interpretation of news will also vary among publications of different types, and among various programmes in the broadcasting services. Each will nevertheless imply some degree of timeliness, interpreted according to the lights of a particular medium.

From the presenter's point of view, news or potential news can be roughly divided into three areas.[7] The simplest is the diary event. It is known in advance that a certain event will take place; the newspaper, the periodical, or the television crew cover the event and report on it. Next we have the unexpected event that must be dealt with. In scientific terms, the unexpectedness can relate not only to the major scientific announcement at a meeting, but also to the ability to distinguish the unexpected from the routine. Lastly, there is what all good journalists, scientific or otherwise, dream about: the successful piece of investigative journalism. Paradoxically, this is usually easier in scientific journalism, since plentiful scientific advice is available, and, at least in the English-speaking countries, the community of science is not monolithic.

In theory, news and opinions are kept in securely differentiated compartments. In reality, the very selection of news items imposes an opinion on the presentation, and most writers will admit to bias. Indeed, a completely objective treatment would be impossible, and, provided the bias of the writer is known, it need not interfere with the transmission of information. The difficulty arises when scientific news is fitted into the traditional concepts of news transmission. As will be argued later, a great deal of science is news only in its consequential terms, can be appreciated only in particular contexts, and has to be translated into a new language, with the result that the old certainties of written or broadcast journalism cannot be relied on.

According to the self-professed level of the medium, there are therefore a variety of approaches to scientific communications. At the truly popular end of the scale, scientific and technical facts hardly exist unconnected with some human, more often that not sexual, freakishness or misdemeanor. Interest is focused almost entirely on a personal aspect, and one need not be over-afraid of appearing patronizing if one suggests that the general intellectual content could be hidden under a very small pebble. Progressing towards those newspapers, weeklies, monthlies, and television programmes whose readers or viewers are assumed to have not only a mind but a desire to exercise it, the intellectual landscape becomes more interesting, if more confused. Here the eternal dilemma of the scientific communicator reasserts itself with renewed force: should an article be about the results of a scientific study, about its (usually complicated) theoretical background, about its possible effects (which inevitably results in speculation), or some uneasy mixture of all three? How far should he make obeisance to the great god News, especially when in this context news can frequently mean no more than the latest undigested information that the editor or the features editor picked up at last night's party. The results are compromises, hardly ever satisfactory either to publication or science correspondent. Examples range from the largely unreadable snippets in the *Nature–Times* news service, which fall neatly between two stools—being far too complex for the general reader and irritatingly superficial for the professional scientist—to the highly polished and professionally rewritten articles in *Scientific American*, which unashamedly require a fair degree of general scientific knowledge. Unfortunately, hanging over all these endeavours is the realization that the seven lean years are now upon us, as witnessed by the death of scientific magazines and the precarious health, with very few exceptions, of the remainder. Editors are not paid to be altruistic, and they will not sacrifice the economic well-being of a publication for the dubious honour of championing science for the people; the total amount of science writing is currently decreasing, despite the occasional flashes of illumination.

In terms of the media scientific news should be timely and convey, implicitly or explicitly, a view of a social or human problem. Ideally the communicator should be a scientist with a deep knowledge, not only of all the facts which constitute the news item, but also of its significance in other spheres. It is a minor tragedy of this century that, with a few exceptions, scientists have not taken on this task. Their understandable, if regrettable, reasons reside in their educational, professional, and cultural backgrounds.

In nominating representatives of the genus scientist, the media must rely on criteria other than the purely scientific. This is reasonable enough, since they lack the necessary yardsticks of scientific excellence. Thus the representatives are chosen usually on the basis of their worldly success and by their appearance which conjures up an aura of expertise. The method works surprisingly well, since, with notable exceptions, the first-rate scientist does tend to rise to the top after the passage of years. But it does mean that the scientists in the public eye are eminent and therefore middle-aged. It has been shown fairly conclusively that an educational revolution takes 50 years to complete.[8] The eminent, the middle-aged, have been brought up on the educational ideas current thirty years ago, which in turn were first developed and put into practice about fifty years ago. As a result, at any given time the spokesmen for science mirror ideas they were imbued with during their adolescence, representing the common wisdom of a period fifty years past, when, in Anglo-Saxon countries at least, true science was regarded as a close approximation to a monastic existence. The participation of the scientist at the level of public communications, as distinct from his presence on committees of various degrees of secretiveness, was regarded not so much as wrong but as simply not done. Such a state of affairs can be correlated with the economics of science at the time: negligible support from government and little more from industry. As the funds were minute, the gentlemanly lack of professional in-fighting was sustainable. Today the situation has changed, and because research funds are immensely larger, we find a great deal of social and professional approval by his peers for the scientist who descends—or perhaps ascends—the political boxing ring. Yet, despite current endeavours, the scientist indulging in the communication of either scientific results or the ideas of science is greatly handicapped by his educational upbringing. With very few exceptions, he is utterly unable to place himself on the side of those who are scientifically not only uneducated but actively inimical. The scientist appears to be unaware that the need is not for a major dilution of ideas, but rather for a translation. Most non-scientists have a not unreasonable objection to being treated as idiots simply because of their unfamiliarity with the vocabulary. It is only recently that scientists have discovered with horror that particular words in one discipline may have completely different connotations in another, and this problem is far greater when translating the ideas of science into non-scientific terms. It has been said that the greatest barrier between America and England is the superficial identity of language, and in the same way the word 'energy' may conjure up completely different images for a physicist and an historian. The lack of a common vocabulary is exacerbated by a *cordon sanitaire* drawn today around the logic and internal culture of science. The automatic assumptions and folk myths of science are no longer contained in the greater myths of the outside world, and thus any meeting between the two will inevitably cause friction. Because of these influences, which basically have nothing to do with the actual or imagined difficulties of scientific fact or theorem, the dialogue between scientist and non-scientist cannot be satisfactorily dictated by the scientists themselves.

Information processing

The theme of this book is an examination of the ways scientific news is, or can be, processed for the enjoyment—or at least consumption—by people other than scientists. In this context the role of the science communicator, who for the sake of simplicity will be referred to as a science journalist, will also be examined.

News about science

There are a number of factors that define the final shape and content of a science item, not least the readership of a particular publication, or more appropriately the editor's belief in the interests of that readership or the equivalent considerations in the broadcasting media. Despite the differences between the media, there are also common factors underlying the treatment of science.

The first and most important of these is that the language of science must be translated. The task of translation is quite different from that of dilution or of explanation, since the very syntax of the language has to be changed. Thus it is rarely satisfactory or pleasing merely to explain the meaning of a scientific term, or to draw rough parallels in terms of an everyday action or contraption. For the translation to be successful, the internal logic of science must be presented in a way that is both attractive and accessible to the reader. One can therefore argue that the seeming difficulty of the scientific material is only one of the problems that the scientific interpreter has to cope with and, conversely, that popular science need not be easy reading in the sense that a romantic novel is.

Translation requires a knowledge of and sympathy for the workings of scientific logic and implies the ability to design patterns parallel to those of science. It is only by showing a pattern of thought or action that the importance of a scientific advance can be placed in correct perspective. The pattern, what Kuhn calls the paradigm, is implicit or explicit in the work of the scientist.[9] For the non-scientist it has to be made explicit if some understanding is to be reached. The design of patterns, by showing background and perspective, must also be accompanied by an attempt at showing future importance. This can be scientific, technological, or social, and quite often all three. Ideally, the science story must deal with all these aspects since the non-scientist will be able to assess the interest or significance of the item only from its possible consequential results. In terms of communicating with scientists, the primary sources for the science journalist, this aspect is often the most difficult. Most scientists are, not unnaturally, weary of assessing the future importance of their work in areas outside their expertise and accuse the science journalist of sensationalism or a scant regard for the limitations the current state of the art imposes on their discovery. Although understandable, such an attitude is rarely helpful to scientist, journalist, or public, and it is the touchstone of the best science communicators to be able to overcome scientific diffidence without arousing professorial wrath.

Even assuming that the task of translation and the defining of present and future importance has been carried out successfully, there is a further and most important ingredient. One can define this as the ability to establish a nexus with the reader or listener, of generating some sense of personal participation. The generality of this idea can perhaps be best illustrated by considering the effect on the reader of an excellent novel. Beyond the feeling of aesthetic satisfaction, literary elegance, and stylistic felicities, there is also a sensation of identification either with a character, a place or even an ambience. There is therefore some way in which the reader can give something of himself. Carrying the parallel a stage further, one can analyse the effect of a painting on the untutored observer. Clearly there are a number of aesthetic and psychological factors at work, and the viewer is offering up some part of himself as well as receiving a stimulus.[10] One must argue that the very best science writing does succeed in evoking this response in the reader: a feeling of working together to solve an important problem. The satisfaction the reader derives will therefore be not only intellectual but also emotional— very far removed from a bored and boring listing of a complex of unrelated facts, however well explained.

One must finally require another desideratum for successful and effective scientific communications: they should be seen to be a part of the reader's cultural heritage. The cultural expressions of the arts and the humanities possess their own logic and language, arguably no less difficult than those of science. Yet the difficulties are accepted because they are seen to be a necessary concomitant to the appreciation of the subject. One can see few logical reasons why the same consideration should not apply to the communications of science, and why it should not be underlining the writings of all science journalists. Actual practice is still a very long way from the ideal, which perhaps accounts for some of the tribulations of science journalists.

The paragon who has to turn these precepts into action is the science journalist. Who he (or very rarely she) is has up to now been generally left to the chance operation of editorial dictat, random vacancies, and the suitability of university graduates in the arts, science, or journalism. That matters are in a far from satisfactory state has been shown by the science journalists themselves, who after lengthly deliberations came to the conclusion that it is virtually impossible to define the education and *modus operandi* of a good science journalist.[11]

Among the welter of arguments, one fact does stand out: whatever professional motivations may exist, an over-active love of money is not one of them. With one or two signal exceptions, most science correspondents receive salaries comparable with those of a university lecturer or a fairly junior manager in industry, thus money is unlikely to be among the mainsprings of a science journalist's motivation. There must, indeed, be something very unusual about somebody who wishes to go on public record on a topic which he may not fully understand, offering it to recipients who are basically uncaring. Yet the process does take place, quite often with great success. The paradox may be illuminated, if not explained, by a suggestion that the go-between in science, as in sex, is often respected, if not loved, by both parties. For the science journalist, the winning of respect starts with some endeavour to familiarize himself not so much with the facts as with the thought-processes of science. If he wishes to be elegant and perhaps obscure, he refers to 'the paradigms of science'.

Most science journalists have had some form of scientific training at university, although this of necessity was fairly specialized. Thus, although the facts they learned have limited uses, it is probable that some of the ideas and attitudes of science have rubbed off. Such knowledge, though, is only the mere beginning, for no-one has yet been able to develop a way of teaching the art, craft, or profession of scientific journalism. As one would expect, standards are variable, both in understanding and expression, and in that sense the science correspondent may become his own worst enemy. Neither is there any easily recognizable way in which excellence is rewarded and lack of quality punished. The number of practising science writers is minute, their career structures chaotic or non-existent. Yet currently every advertisement calling for a junior editorial assistant will have up to a hundred graduates applying. Either they must all be mad, extremely foolish, and unworldly, or the task must have some hidden compensations.

The science journalist cannot present himself to scientists of various disciplines as an expert in science. He must instead win their confidence as the representative of another comparable profession, who has the power to collaborate but will not do harm scientifically. He must be a diplomat and a persuader in suggesting that his role in the interpretation of science is beneficial, not only to the public but also to the individual scientist. In his dealings with scientists he must be no mean student of psychology, since by and large, and despite popular belief, scientists are not emotionless.

Through all his work, the science journalist treads a tightrope between the concerns of the public and those of the scientist, especially when, as it frequently happens, there is a genuine and sincerely felt antithesis. He must be able to criticize without losing confidence and to praise without sounding patronizing. All he writes must be presented in a style that is clear, readable, and, like good design, unobtrusive. He must basically be a specialist educator working by stealth, for the public has little desire to be educated, nor does the scientist wish to have his shortcomings revealed; yet without a committment to such an educative process the very foundations of science communications collapse.

But above all the science journalist must have a commitment to science: an open-eyed, realistic, and perhaps even cynical commitment which does not try to hide the human, institutional, and theoretical difficulties of science and the misdemeanours of scientists. On the other hand, he must believe that in their totality the aims, ideas, thought-processes, and motivations incorporated in science are on balance beneficial to the world at large (and not only to the personal ego of scientists, the activities of industry, or manipulation by political masters) and that in the last analysis the aims of science must be congruent with those of a cultured society. If he does not believe all this, although the belief may be only half-conscious, his work is pre-judged to be sterile, and his efforts will be in vain, not only for the two publics he serves but for himself. He will be destroyed not only as an expert and a craftsman, but also as a human being.

2
Is science too difficult?

The difficulty of communicating a facet of science outside the fraternity of specialists in the field has often been contended by pointing out the complexity of modern science: the abundance of facts that even a three-year university course cannot encompass, the mathematical sophistication and conceptual subtleties threatening the trepid enquirer with a substituted steroid or a black hole in the sky. These arguments are reassuring for both inarticulate scientist and unthinking public. If by its very nature the essence of science is untranslatable, why go to all the trouble of attempting the impossible? Such attitudes also bring in their wake a degree of exclusivity. It is implicit in the difficulty of science that its practitioners must partake of very special intellectual skills while, for the non-scientist, pride in ignorance emphasizes that he belongs to real humanity.

That matters have progressed to this state is a comment on our educational system and the attitudes it engenders, on the blindness of both the status-seeking scientist and the unthinking humanist who takes pride in cutting himself off from a major cultural activity of contemporary society. These attitudes incorporate two fallacies. Even a few moments of thought should inform both the scientific and the anti-scientific xenophobe that talking about science does not and should not require the acquisition of large quantities of scientific fact. Even a scientist specializing in one discipline would not claim knowledge of more than a minute fraction of the facts available within the discipline itself; to test the truth of this proposition it is sufficient to listen to a conversation between an organic and a physical chemist and to note that, although quite often they use the same words, the meanings they attach to them are subtly different.

It would be possible to enlarge this argument by questioning the very concept of fact as used by scientists.[12] From the science journalist's point of view the most important problems are suggested by the interpretation of facts leading to their potential consequences. To consider one example, an assessment of the potential dangers of a nuclear reactor is based on a relatively small number of facts interpreted through a massive statistical analysis. The non-scientist will be concerned only with the final interpretation; without an intimate knowledge of metallurgy, engineering, chemistry, and statistics he cannot sensibly comment on the underlying scientific work. The task of the science writer is to place in perspective the processes used to derive the final interpretation, to bring out any major assumptions made, and to note conflicting results if these exist.

Thus in writing about science for those who are not scientists one is not concerned with more than the bare minimum of facts. On the contrary, one should attempt to give an insight into modes of pattern-making, or how simple statements of fact are manipulated with accuracy, elegance, and imagination into a cohesive pattern—which is what science is all about. One of the often-quoted university stories concerns the aspiring young chemist whose school has acquainted him with the jargon, if not with the true meaning, of Russell–Saunders coupling, a fairly high-powered mathematical treatment of certain aspects of cohesion between atoms, but who thinks that sodium chloride (common salt) is a green gas.[13] Although from the harassed university lecturer's point of view it is perhaps unfortunate that he has to group his charges according to their belief in the immanence of green gases, from society's point of view this is hardly as serious as all that. Given that the vast majority of those who had some brush with science at whatever level will not be following a scientific career, their knowledge of scientific fact is immaterial; they can if necessary always obtain it from the scientific adviser on tap. It is unlikely that a Cabinet minister, a judge, or the managing director of a large company—or indeed a milkman or a dustman—will be greatly troubled by his inability to reel off the names of two dozen green gases. One can, however, suggest that at least some of these people will derive great benefit from an idea, however tenuous, of the processes whereby facts have been obtained and evaluated, if only to be able to question the *ex cathedra* statements of their scientific advisers.

As sciences develop, although their factual content increases, their patterns also become more established, so that, once the internal logic of the science is acquired, only a very few facts suffice. But, it is argued, the maturing of a discipline is accompanied by an increasing complexity and sophistication of its underlying theories, invariably requiring a

Is science too difficult?

knowledge of mathematics. The theories of today's physics are far more difficult that those of Newton's; the modern chemist takes as naturally to the computer as his predecessor to the Bunsen burner. Consideration of the nature of scientific law and hypothesis would extend the discussion far beyond the scope of this book, yet it is perhaps not irrelevant to note that even among scientists and philosophers of science there exists little agreement on the nature of these concepts. As far as the non-scientist is concerned, one can cite Dainton's observation that the meaning of scientific law may be considered in two ways: as an answer to questions asking how, or to questions asking why.[14] It is not necessary for the driver of a motor-car to be familiar with the chemistry and thermodynamics of combustion, or with the corrosion resistance or hysteresis properties of steels. It is sufficient for him to be aware of the existence of the internal combustion engine, to realize its power and perhaps the existence of exhaust fumes. One can therefore suggest that precisely in the communications of science answering the 'how' question is much more important than dissertations about 'why' which properly fall within the ambit of the specialist. On this basis it is possible to assert that theories communicated need not incorporate mathematical formulations, and a simplified, qualitative description can generally be provided. Such an explanation will be sufficient to explain the behaviour of physical entities, although it will not provide an insight into the underlying mechanisms.

At this level, it is possible to argue therefore that, for example, chemistry, where a few major theories can explain a vast gamut of data, must be a considerably easier science to learn today than in the time of the alchemists, or indeed up to the middle of this century, when an army of apparently uncorrelated data was apt to descend on the head of the unsuspecting student. Botany before Linneus must have been torture, but a systematic exposition of groups and subgroups can assure us that the discipline has not been invented solely for the discomfiture of students.

We must also consider the educational basis of scientific studies, amateur or professional. Information about science does not impinge on an untutored mind, but one that has partaken of whatever education its times and its society offers. Thus the difficulty of science must be considered in relation to the general educational level of a particular epoch and of a particular class: it is a relative rather than an absolute problem. If one considers the educational background upon which any scientific communications must have been imposed at various times in history, it is difficult to escape the conclusion that in any age the relative difficulty of learning about science must have been at least comparable.

Is science too difficult?

The amount of mental effort needed to become familiar with the ideas of contemporary science must always have been considerable.

Yet people did take the trouble. We must therefore assert that in communicating science we are dealing mainly with attitudes: attitudes of society, of education, of faith, of social imperatives. We are dealing with what particular segments of society regard as correct and what they affect to despise. We are dealing with professional ethics and popular mythology, and also with economic realities and their effects on people. We are not dealing with intellectual difficulties.

In looking back to the days when every educated person could claim acquaintance with the science of his day, the amateur dabbling in science constituted a proof of culture and gentility, and contrasting them with the brutish crowds of today caring more for Coronation Street than chemistry, we ought to beware of matching like with unlike. It is doubtful if the seventeenth-century equivalent of a *Daily Mirror* reader cared a great deal about science in the sense that a *New Scientist* reader does, although he might have been superlatively gifted at some technology of his day in the same way that our contemporary radio constructors accomplish their complex tasks without any awareness of solid-state physics. We cannot therefore define the public towards whom any scientific communication or education might have been directed. There must be more than a vague suspicion that, taking one decade with the next, one is really discussing educational and social distinctions, and thus by implication particular constellations of power politics in England.

Up to about the middle of the nineteenth century, a man of education would have considered a knowledge of some science or mathematics to be a reasonable accomplishment. How far scientific knowledge permeated the social structure is debatable and cannot form part of the present argument, but it is interesting to note that in an age when most science was written in Latin, Galileo chose to make his statements in Italian, and not surprisingly was not universally applauded. By the seventeenth century in England we see the rise of Puritan values and the birth of the new science, which no longer depends on Aristotle's and others' *obiter dicta* from antiquity; science was becoming particular and specific instead of universalist and general, and the Industrial Revolution started its gestation period.[15] There have been a number of attempts to correlate the Puritan ethic and the advent of the turmoil of inventiveness with its social and economic effects which go under the name of the Industrial Revolution: the fact remains that knowledge of the new science was spreading within the social structure, for the new race of inventors was separate from the academic establishment and the aristocratic amateur. How much relationship there was between the theorists of science and the practical men is debatable, despite Ashby's statement that 'there were a few cultivators of science engaged in research, but their work was not regarded as having much bearing on education and still less on technology. There was practically no exchange of ideas between the scientists and the designers of industrial processes'.[16] The foundation of the Royal Society was part of a movement rather than a unique event; at one time in the eighteenth century there were thirty flourishing local scientific societies, the Lunar Society of Birmingham and the Literary and Philosophical Society of Manchester being the best known.[17] These societies, the Royal being paramount, regarded it as their duty to further not so much science as the useful arts; in other words the motivation was economic pull rather than scientific push. Within this framework, they regarded the popularization of science and technology as very much part of their task. Yet they failed.

By the 1800s the Lunar Society had died and the Royal Society had fallen on bad times. Britain's technological lead over the foreign competitors was gradually being lost, a condition that was to become apparent by the second half of the century. This period, too, was taken up by a great many arguments over the place of science in the educational system, and the relationships between state aid and scientific innovation.[18,19] Whether the growing supremacy of German chemistry was due to a greater abundance of specialists, or to the good second-raters— what today would be called generalists—or to the better co-operation between science, industry, and the state were matters of urgent debate. One would suspect that the debate had few practical consequences, since at the outbreak of the First World War large sections of the British chemical industry had to be created to supply materials previously imported from Germany.

During the nineteenth century, the proponents of science may argue, the great popularizers —the Faradays, Huxleys, and Tyndalls— collected their admirers, and *The Times* could publish a lengthy report on the discovery of noble gases by Ramsey and Rayleigh;[20] but an undoubted chill in the air was becoming noticeable, and science was becoming unfashionable. The rising public schools regarded science as suitable only for those who did not show promise for the Church or the Civil Service, an attitude that lasted well into this century. The formation of the mechanics institutes, which became the latter-day polytechnics, was no counterbalance, for they spoke to a different segment of society, a segment which then mattered little in political and economic terms. Perhaps the most notable indicators of the changing general view were the poets, crystallizing and articulating the unformed public view:

Keats, who had six years' medical training. Shelley the chemist; Coleridge, Tennyson, Wordsworth—all at some time took an interest in the science of their day, only to turn away dissatisfied and uninterested. The problem was stated clearly enough by Wordsworth: 'The remotest discoveries of the Chemist, the Botanist or the Mineralogist will be as proper objects of the Poets' art as any upon which it can be employed, if the time should ever come when these things be familiar to us.'

By the end of the nineteenth century public concern with science seems to have crystallized into a familiar pattern. Those aspects of science which have an obvious bearing on spiritual or material interest, which could establish contact with the real or imagined feelings of the man in the street, were in the public arena. The great Darwinian controversies and heroic tales of explorations were easily accommodated within popular communications of science. For the physical sciences, where only their second- or third-hand consequences could be seen, the romantic poets may have voiced the true public opinion. Perhaps a whole nation became so appalled by the excesses of the Industrial Revolution that it decided not to have anything to do with them. This is not a compelling argument.

It might have been an understandable emotional reaction, yet intelligence must have informed the protagonists that, rather as today, given the despoilation of applied science, only science itself could put matters right. One is inevitably driven to the conclusion that the causes of the anti-science movement were fundamentally social and that, the cycle of wonders of science having been exhausted, there was nothing left to take its place. In a society of two nations, those who worked the dark satanic mills had no reason to be interested in science nor had they energy left for it. The industrialists cared little, and the conservatism of British industry became a byword—a conservatism whose consequences have reached through the century to our own days. Neither did the governing classes have much need for or interest in science. With the slackening of the economic pull, no intellectual push presented itself to propel science into the public imagination.

Decline in the public interest in science can thus be roughly dated from the second half of the nineteenth century, bearing in mind that one is offering a rough generalization rather than an accurate analysis. Despite a number of exceptions, despite the great outburst of interest following the Second World War, a lack of interest in the sciences is not a new phenomenon, although the increasing power of science and technology endows this problem with contemporary urgency. The

Is science too difficult?

communication of science to the general public is not, however much we would prefer it otherwise, a function of the difficulty of contemporary science relative to some arbitrary educational level, neither is it tied to the importance of science, its intellectual beauty, or the skill of its interpreters, although all these factors naturally contribute. Basically, one must argue, success in communicating science is involved in a complex but very real way in the real or imagined needs of society or, more precisely, of those groups in society whose views command power and influence.

If scientists or others are truly desirous that society should be accurately aware of the contributions that science is able to make, they should not concentrate their fire so much on education or the widest possible dissemination of scientific ideas. They should rather attempt to exert their influence on the ruling strata of society, whichever they may be. It is perhaps no coincidence that countries under a centralized communist or—as they prefer to call it—socialist government, place great emphasis on the popularization of science. It terms of popular scientific output, Russia is by far the largest producer and consumer today. Neither is it by chance that precisely in these countries the ruling ideology asserts the supremacy of the scientific method, at least the scientific method as put forward by the founders of the system. One should emphasize that such an argument does not attempt to relate communications to the quality or quantity of science in these countries, a much more complex and historically determined matter; it merely asserts the need for a strong power base, in terms of society, for the full operation of scientific communications outside the sphere of specialists.

The institutions of science

To what extent should the scientists' own institutions be involved in communicating science outside their disciplines? The answer is by no means clear-cut. Since its foundation in 1832, the British Association for the Advancement of Science clearly thought in the affirmative. Its founders had some very clear ideas on what the new body might do, and they lost no time in attempting to obtain the support of leading scientists of the day; there is a long correspondence in Faraday's letters about the formation of the Association. For example, Vernon Harcourt thought that the new association, which was to spring from a federation of existing scientific societies, should encourage scientific research, especially in new areas and should become the centre of a scientists' lobby which would influence the government of the day to obtain better conditions.[21] By 1832 the British Association was holding its first meeting in Oxford, and the list of those attending subsequent meetings reads like—and is—a list of all the most eminent scientists of the day, both from Britain and the Continent. Yet already the Association was losing its way, despite its unarguably great success, the well-attended meetings, the reading of papers disclosing new discoveries. The British Association became, paradoxically enough, too scientific. The faithful came every year to hear the lectures and between times they read the *Journal*. With the growth of big or even medium-sized science, the Association lost one of its planks; what it did or failed to do had very little bearing on the course of science, and its advocacy for *scientists* in politics was quietly forgotten. Even today, the British Association has not more than 6000 members, a pitifully small number for a body that claims to speak for the whole of science and for all its practitioners. A part of this failure is that of organization. There was a long and sad history of personal and functional ineptitudes, culminating in a great public debate and the setting-up of a commission in 1972 to look into the future of the Association. The Commission's recommendations were fundamentally an advocacy for returning to the old roots and moving to a far more popularizing stance.

In addition to its traditional activities centring round the annual meeting, the B.A. is also deeply involved in scientific activities of schoolchildren—the British Association of Young Scientists—and university students—the British Association Student Section. As for communicating science, Magnus Pyke, Secretary and Chairman of the Council, wrote: 'those who understand ... scientific principles have a great responsibility to explain what the facts are and what their social consequences may be.'[22] Thus the B.A. provides a forum for a number of *ad hoc* study groups to report on matters of the moment, for example the group on Social Concern and Biological Advances. The stated aim of the Association, to provide journalists with a locus for obtaining (scientific) facts, and indeed to act as the Peripatetic University for the enlargement of scholarship, is still unfortunately more in the realm of hopes than in reality.

The troubles of the British Association are only a specimen of the difficulties being visited on scientific societies in most of today's developed countries. The high birth-rate of learned societies of the type we know today took place in England from the late 1840s onward, in the United States somewhat later. In almost every case the society was

Is science too difficult?

to occupy itself with only one discipline, but within that framework was to encourage not only science but also its application, and by implication its popularization.[23] In certain instances, as in the case of the Royal Institute of Chemistry, the concern with public weal was explicitly written into its constitution. Economic evolution having culled the unviable, the surviving societies guarded science well, scientists less well, and public knowledge and application not at all. The reasons are written large within the very constitutions: once concern with science itself had become an irresistable pull, the rest had to follow. The movers of societies would be eminent scientists, or scientists hoping to be eminent through leadership of the society. Conversely, public concern was overshadowed by science, even as far as their own members were concerned.

Today's respectable and well-run society will typically consist largely of a publishing enterprise for one or more research journals, together with a good library and the skilful organization of more-or-less specialized conferences. The application of science and the education of the public are more apparent in statement than in practice. Ironically, the fate of big science has overtaken the societies. No longer will the success of a particular piece of research depend on the professional approval of Council, but rather on basically political decisions by government or its agencies. Neither will communications from the cutting edge of science depend on the reading of original papers at meetings organized by a society; they will be published after a decent delay of up to nine months and will take their place in the great river of sometimes important but often trivial information, to the great delight of computerized information retrieval men. The public, even in the shape of a society's own fellowship (for are they not also members of society) show muted concern, a suspicion that something ought to be done, without a very clear idea what the something should be. Although the apparatus exists and committees meet every day declaring that only science and technology can help to extricate mankind from its present morass, the gulf between learned societies and the mass or even the semipopular media has become too vast for substantial bridging to take place.

A number of ultra-respectable, not to say ultra-conservative bodies such as UNESCO have their own pet revolutionaries who can be relied on to accept invitations to meetings and give the standard critique of science address, prefacing it with expressions of their high concern for the future of science and the liberality of the hosts. Nevertheless, one must assert that one hippy does not make public concern, and in the final analysis the societies realize that in future their concern must go out to their membership, to the scientist rather than to science, if they are not to end their days as large specialized publishing companies making their bread and butter on the library subscription. The learned societies, which started life as clubs for mutual encouragement, have now to face their own contradictory roles. The plant they so carefully nourished in their own private greenhouse has shot through the roof and threatens to engulf not only them but all in sight.

Communicating science to the general public could have formed part of the scientific societies' role had they so chosen. Most of them possess education departments whose role is to increase the quantity and quality of teaching of the discipline, usually at school level. An outstanding example has been the American Chemical Society's contributions to the Chem Study project, which revolutionized chemistry teaching in American schools and had an enormous influence in other disciplines. But this is communicating about science at second hand. Faced directly with public attitudes, the societies have done little beyond the appointment of a press officer to make abstracts available at the annual meeting and to attempt to manoeuvre the local television station into granting some scientific luminary a five-minute spot in a magazine programme. Nor is the task as simple as it would appear. The societies are in a state of transition between caring for the science and caring for the scientist; and the aspirations of the scientists, as expressed in professional terms, are not automatically congruent with what the public considers to be its best interest. In the past the societies often expressed horror at the thought of indulging in public relations. Although such ideas are now considered to be outmoded, it is still true that popularization is looked upon with disapproving tolerance, perhaps cloaking a feeling of impotence faced with a seemingly easy but actually very difficult task.

An interesting exception to the run of learned societies has been in Britain the Royal Institution, founded in the last year of the eighteenth century. Throughout its existence the RI has always been a research institution, running its own laboratories and thus partaking of both the successes and tribulations of real science. It also has a strong tradition of the lecture demonstration, and, although during its history it had its off-colour periods, this tradition has always been followed. The Friday night conversazione, when an eminent scientist talks to a lay audience about the state of the art, has been and is regarded as very important, as are the lectures developed for schoolchildren. Thus the RI has shown how a learned society can be involved in the popular communications of

Is science too difficult?

science. Yet its membership is less than 2000. The unkind thought must occur to the observer comparing the popularizing work of the RI with the lack of activity of its sister institutions that, although the former might have been founded by an adventurer, its social sights were never in doubt; the founders of the RI meant to influence those who mattered—the aristocracy—and neither they nor the recipients of their information were worried by the intrusion of vigorous showmanship. Whereas the learned societies are firmly middle-class foundations, spending a significant part of their time looking over their shoulders to see if they remain soberly dressed and respectable—which they do—and which hardly encourages them to feel and act enterprising and imaginative. As a unique prototype the Royal Institution has therefore proved to be an exception to all the rules and arguments about scientific communications. Its scientists—or at least a significant number of them—can communicate; it finds a ready audience for its communications; and at the time of its foundation it possessed a clear and direct idea about the need for communication in relation to the power structure of contemporary society. Under present conditions the work of the RI still goes on (and may it long continue), but the structure of society supporting it has changed, and the infrequent presence of a duke at a symposium will make little change in the structure of scientific endeavour in the future. The RI has fallen a victim of the horrifying growth of the system; in early nineteenth-century London those who might be interested in science and whose influence was worth moulding could probably be accommodated in the lecture theatre to hear a discourse by a Davy or a Faraday. Today only a rare televised occasion can hope to make a comparable impact.

Underlying all attempts at popularizing science must be the contemporary state of education, in the effective rather than the formal sense. It matters little if during x years at school a pupil has y hours of various scientific disciplines thrown at him, or, in the American terminology, is exposed to them. It is, however, important what remains at the end of his schooldays; whether ten years after the finish of the educational process he still retains enough knowledge and interest for any further scientific information to have significance in his eyes. It is straying too far from the ground of popular communication of science to enquire deeply into the history of scientific education in Britain, either at the secondary or at the tertiary level. One can note all the same that most of the current arguments and a great number of the educational experiments have already been tried, at least on a small scale, as far back as the early decades of this century. The heuristic method in the schools was at least theoretically known; the great arguments about specialization at university level were foreshadowed by equally emotional diatribes when University College London was founded in the 1830s and re-emphasized later in the nineteenth century when government and academics had a fine series of rows about the place of science in the curriculum.[18]

Since the Second World War the truly appalling state of scientific education has given rise both in this country and in the United States to the Nuffield-type educational technique and ideas, but it must be said that only about 10 per cent of schools in Britain have actually adopted a Nuffield approach.[24] Even so, it would be pleasing to suggest that, given the sudden improvement in educational methods, there appears to be a resurgence in popular interest in science emanating from all those finely honed, discovery-through-experiment-tutored minds. There is absolutely no evidence that this is so.[25]

It seems clear enough that, at least in English-speaking countries, the science departments of the universities are suffering from a lack of undergraduates. The situation in the secondary schools is more difficult to interpret as part of a long-term trend, although there are indications that at least up to the British A-level there still is a fair amount of interest in science.

Similarly, we are too prone to assume that public disquiet with science will express itself in direct disapproval or hostility. This does not appear to be so. For example, it has been shown that confidence in those leading the scientific community is still high in the United States, and furthermore that it has not changed perceptibly in the years 1966–73.[26] 'Science seems to command more influence than many other institutions', is the comment made by Etzioni and Nunn. Yet 'The [science] writers perceived that public attitudes today were far more negative than they were in 1965, and they believed that the Vietnam war and environmental concerns were the two elements most responsible', was the opinion of S. M. Friedman, analysing a survey of American science journalists.[27] The situation is still as before: those to whom science is attractive, either as a vocation or as a system of ideas, become scientists. They may not become good scientists, but at least practice and convention will ensure that they pay some interest to the affairs of science. For the rest, whether the educational process took place in the public or the private sector, at whatever degree of privilege or lack of it, the non-scientist is firmly and quietly uninterested in science and all its works. Not because the facts or ideas or idioms of science are beyond him intellectually, but because—and this seems to be emotionally difficult for scientists to accept—he is simply not interested in science. In this respect, the average member of the public appears to be little different from his predecessor, at least during the last two centuries.

Attempts at popularizing science have been a true expression of real or imagined popular attitudes. In the seventeenth century, the general tenor emphasized the utilitarian aspects of science and technology; the development of better agriculture, the creation of wealth were taken as seriously as the philosophical assertions of science about the nature of man or of reality. In the nineteenth century we begin to see another important trend: science as a form of play, the delight in the experiment that can be performed, coupled with its generally didactic tone. Here again, one can observe the phenomenon of personalization that tends to be so berated by serious scientists today. The popular Victorian book or magazine article on science will more often than not describe an experiment or a newly found morsel of knowledge in terms of what a particular scientist did or found. Even scientific papers, which today are the formalized assertions of a faultless and largely imaginary logical sequence, then still maintained an aura of human endeavour where hopes and disappointments were of equal weight to discovery.

Also during the Victorian period emerges the last strand in science communications: the idea of social concern, of relevance of science and technology to the fabric of society, a trend that was to become so powerful after the Second World War. There were a number of worthy successors; the works of Haldane, Hogben, and Russell earlier this century have sought to place science in the context of society's needs and activities, but it is arguable that they did not reach those classes of society for whom the word popularization is perhaps used with most justification.

In a sense the indicator of societal concerns is the science writer, who by dint of his own preoccupations magnifies and sometimes catalyses such a concern. We shall argue subsequently that newspapers and periodicals, by their own submission, cannot alter the general feelings of society, but they can, by the very act of articulating them, help to magnify and bring to fruition an already existing tendency. On this assumption, it is reasonable that in the post-war years preoccupations were centred round the marvels of science or, more accurately technology, then coming into use which seemed to promise the preservation of peace and the advent of plenty for mankind.[29] By hindsight, this was a simplistic view, for it omitted reference to a number of deep-seated human drives and emotions whose gratification can ultimately be achieved only through political and social channels. The Vietnam War and the growing concern with amenity voicing itself through ecological and environmental polemics have ushered in the current state of science reporting. Now, while the discussion of technology is still in an essentially Victorian form, consisting of the description of a discovery and its utilitarian value, science reporting has become firmly grounded in the concerns of society. Few books, articles, or television programmes can be presented today dealing with some aspect of science which do not attempt to relate it to the individual either in his own or his societal capacity.

One can argue that the current obsession is just as bad as the unthinking eulogy of science in the past. Such an argument has great substance and is at the same time irrelevant. Naturally enough, a given social climate will bring forth advocates on its behalf, and it would be unreasonable to expect them not to exaggerate the burden of their arguments. Once a bandwagon starts to roll, human nature will ensure that not only the less disinterested but also the less skilful will seek to climb aboard. There must be fashions in the reporting of science—for scientists as much as for the general public—which at the time appear to be random but, after the passage of time, will be found to have captured some of the spirit of the age to which they belong. Currently, scientists complain that science writing emphasizes the negative aspects of science and gives no comfort to those who see in science and technology the solution of all our problems. It is odd that they should be so surprised, and odder still that it was to a large measure due to the science writers that the problems science created in society came to be recognized. Indeed, scientists should feel pleased that there is still enough emotion clinging to science that some feel strongly enough to commit themselves, for the truly great tragedy for science will arrive when nobody feels passionately enough about the affairs of science to attack them. Excesses will be committed, scientists misreported, conclusions taken to absurd lengths—but do they matter? If there is to be a general argument about science and its role in society, somebody has to start, and, as I have tried to argue, the scientists, with very rare exceptions, are singularly unfitted to do this.

Let scientists relax; the public gets the science correspondent it deserves. The excesses of today, the apocalyptic vision through the claustrophobia of industrial society, will change and will be the subject of innumerable PhD theses in the years to come, like the mini-skirt and

Is science too difficult?

rock-and-roll singers, it will be regarded as an expression of particular humours of our society. Whatever its particular benefits or drawbacks, the science reporting of the late 1960s and early 1970s of this century has at least kept the interest of science and its consequential effects before the eyes of a very tired, disenchanted, and wary public. For that it ought to be given a vote of thanks.

3

Science in the Press

The intuitive image conjured up by the phrase 'communicating science to the public' is of the printed word, and particularly the world of newspapers. In reality the public for television programmes is greater; radio instructs the housebound housewife; books and journals are paramount in the academic and pseudo-academic mind; but by reason of history, self-arrogated traditions, and the tattered laurels of public esteem the Fourth Estate is still king.

The following comments are based on conversations with people writing for and producing newspapers, who regard communications as a job rather than a crusade, and who therefore talk about reality rather than about the desiderata of politicians and sociologists. My comments refer primarily to the national press in Britain, although in most countries the same pressures are evidently at work. In particular, the same problems are beginning to be discussed in the United States.[29]

The presentation of science in the newspapers represents the outcome of a compromise between personal, economic, and political pressures. According to their attitudes to science communications one can divide newspapermen, roughly speaking, into two groups: the science specialists or science correspondents and the rest, represented by editors or members of the newspaper management.

In essence the science correspondent faces a twofold problem. On the one hand, he has to overcome various difficulties concerned with gathering and translating information; he would regard these as professional challenges. On the other hand, he has to be very conscious of working within a particular hierarchy and of representing a world far removed from that of his editors and executives.

Let us assume that the science correspondent has obtained what to him appears an important and newsworthy story. His first task is not explanation, but selling, for he has to persuade his colleagues and superiors on the paper that the story needs to be published. This he has to do not only in a highly competitive atmosphere, but also in a great hurry. In quantitative terms, between half a million and two million words pour into a newspaper every day, of which only about a tenth can be published; and all executive decisions, writing, accepting, and editing must be accomplished in a matter of three of four hours a day, mainly for technical reasons. As David Perlman of the *San Francisco Chronicle* observes, 'The science writer is considered (and sometimes resented) a specialist in a field that demands too many long stories—stories that pose too many unfamiliar concepts...'[30]

The science correspondent is not usually a good salesman, and more often than not he sees his story downgraded into a minor item, if indeed it appears at all. One has an overwhelming impression that to be a good science correspondent it is necessary not only to be able to clothe each lamb in wolf's clothing but, no less important, to establish a sufficiently good working and political relationship with one's colleagues and superiors, so that the inevitable give and take of a daily high-pressure round of friction causes no personal damage. Personalities are therefore just as important as a knowledge of science and its presentation. It would be only a slight exaggeration to suggest that the physical appearance and personal mannerisms of the science correspondent have a great influence on the stories that appear in the paper. The harmonious team in most close-knit organizations is usually a myth, and stories of rows, alienations, and spectacular bitchiness are the staple fare of newspaper gossip.

For the more imaginatively endowed, or for those whose political inclinations or personal idiosyncrasies tend towards the conspiracy theory of human affairs, the simple fact that pint pots can only infrequently carry quarts can be elevated into elaborate social or political critiques of the system, and, since collections of people engaged in roughly the same pursuit must constitute a system, the arguments carry weight. One of the main ones concerns the explicit commercialism of the news media. One disgruntled science correspondent has remarked that if the Queen were to die tomorrow the front pages would be swept clean of any other news. Indeed, but what is even sadder is that the same effect can be accomplished, as far as news of science is concerned, by rumours about the love-life of a pop star. It is hardly surprising that science correspondents have been characterized as basically unhappy men and little weight is given in general to their contributions. The fundamental difficulty, as seen by the science correspondents themselves, is that their papers are fickle and will not allow a thorough (meaning lengthy) analysis of a problem over a sufficient period of time.

Even the great ecological controversy, which had been treated at various levels of intelligence at a fair length by all newspapers, has been allowed to die out of sheer boredom: but boredom on the part of the newspaper editors, and, one suspects, of the reading public, rather than the correspondents.

The energy crisis, together with the sudden uncertainty about the future of all primary resources, has also presented a golden opportunity to bring science, or more appropriately technology, into the forefront of public consciousness. For a limited time, everybody becomes a conservationist, and the inevitable if ineffectual 'I told you so' articles sprout in every newspaper. But since the energy and resources argument has been in existence for a long time without, it appears, anybody at all taking the slightest operational notice of it, one may be permitted a degree of scepticism about the actual influence of scientific communications in this realm. Such a state of affairs can be described in such terms as 'Fleet Street never adjusts to what is happening outside', or by stating that somebody has to decide on the public's interest and in the last resolve that is—or should be—the editor of a newspaper.

This is a commonsense but misleading deduction. For one point of agreement is obvious among the science correspondents, the self-appointed socially relevant critics, and most newspaper executives: that policy, especially long-term policy, simply does not exist in the world of Fleet Street in any sense that would be recognizable in government, a foundation, or a major industrial enterprise. It seems that newspapers, as perhaps they must, exist from day to day, without more than the most tenuous underlying strategy of long-term action. A multiplicity of *ad hoc* decisions are made in the heat of the moment, not infrequently on insufficient evidence and less understanding, but on a great deal of intuitive experience. If a paper is lucky, it will have built up a team which, at least in general, tends to think along the same lines, resulting in a vague consensus, usually gratified with the name of policy. The science correspondent may rail against the inadequacy of the system and the blindness of his editor. In reality the editor is quite unlikely to be in a position to judge, or indeed to see his copy before it is printed; the decisions will have been taken at a much lower level at a far greater speed.

There is also a surprising area of agreement on the adequacy of the sources of science stories that finally appear in a paper. It would be charitable to say that they are random. The science sources of one of our great, respected, and still profitable national newspapers are: one science correspondent and a science editor, who between them are supposed to monitor all the advances of all the sciences; one university physics lecturer who is married to a fairly junior member of the staff; and a second physicist at another college who happened to have been a contemporary at university of the features editor. In Fleet Street terms, this is a pretty good communications network, and indeed virtually every edition of this particular paper carries some item concerned with science.

But the features editor is unhappy; the excitement seems to have gone out of science, and the offerings of his science specialists are grey and uninspiring. To be fair, the editor in question has his focus on higher things. To paraphrase: 'What we want is big, important, glamorous and immediate stories, which are gripping. Let me give an example. When X (he got the name wrong) discovered Y and wrote a book about it, we missed the serialization rights. So I got S (the name of a well-known feature writer) and I hired P (he got the name wrong again—it was a very junior member of the scientific team who emerged from the process completely uncovered with glory) and P gave him lessons. How many? Oh, he came in for three afternoons. And do you know, S got so interested he started writing all those funny formulae and we had to restrain him, but he did write 5000 words and we were all excited.'

This particular discovery involved at least three branches of science and was the culmination of about eighty years of research. It resulted in three Nobel Prizes and an almost completely new scientific discipline. One has therefore to look up in awe at Mr. S and Dr. P, who between them wrote it all up in three afternoons starting from nothing.

What is undeniable is that the flow of news towards the science correspondent is poorly organized. In the survey referred to,[31] the science correspondents on three American papers found that their sources of information included scientific conferences, news releases, talking to scientists, and a random reading of scientific and quasi-scientific magazines.[31] To which it would be reasonable to add the excitement of non-scientific colleagues whose children fall ill, whose gardens are invaded by pests, or whose relaxation is interrupted by aircraft noise—all good scientific stuff that needs looking into.

A recent survey analysed the subject-matter of science stories appearing in three major American newspapers during 1973.[31] In a one-week period the *Los Angeles Times* published 22, the *New York Times* 32, and the *St. Louis Post-Dispatch* 16 science articles, the great majority of them being of less than 20 paragraphs. Between a third and a half of the articles dealt with medicine and public health; between 10 per cent and a quarter with agriculture and biology; while all the others, for example the earth sciences, represented well under 10 per cent each.

The quality of scientific reporting, or rather the lack of it, rumbles through all these conversations. Editors accuse the science writers of being dull, of trying to be pompous and didactic, of acting as a lobby for the scientific establishment, of being insufficiently in touch with readers' tastes. Conversely, the radical critics make the same point, from the other extreme, accusing the science correspondent of being out of touch and insufficiently motivated by society's concerns. Since there is no mechanism whatever to measure the readers' interests except the rise or fall in circulation figures, this is a satisfactory argument for all concerned because nobody can win.

It would be a simple matter to blame the papers for this chaotic state of affairs and to suggest that a small fraction of the moneys used for marketing purposes would be sufficient to provide a proper structure which would encourage the flow of scientific information. Such an argument would cut little ice with individual managements, who without exception blame the scientific community for its inadequate sense of what constitutes news. For example: 'Scientists keep rabbiting on about how badly they are treated, but hardly ever tell reporters about new discoveries, and when they do they are patronizing. In any case, they are inarticulate and cannot communicate results.'

But what results should they communicate? One editor assured me that, should anyone invent a surefire cure for influenza he would get front-page treatment. Otherwise, the products of what he quaintly referred to as the science industry—meaning drugs and mechanical gadgets—must affect either the body or the purse of the average reader. However, the paper concerned did publish a shock issue on pollution.

The need for scientists to explain was a constant accompaniment to all discussions, together with a hurt accusation that they will not be properly questioned by reporters. So what is the paper to do? As one editor pointed out, the newspaper has no duty to scientists, and science being taken for granted can be dealt with only as a background comment in terms of other implicitly more interesting issues. Such a general view of science reporting was by no means the prerogative of the popular papers; those that take pride in their intellectual or social standing expressed the same thoughts more elegantly in somewhat better accents.

One reason for this treatment is the implicit, and often explicit, view that a great deal of writing received from the paper's own science correspondent is uninspired or downright bad, and is being kept on as a

Science in the press

general background fill-in, a token of good intentions. Only one paper in Britain—*The Times*—carries a daily science item outside its own internal lines of communication: the *Nature–Times* science reports, modelled on the Law Reports. It is only fair to note, without comment, that nobody on the newspapers one has talked to found them readable. Yet, whatever the measure of quality adopted by the newspaper world, a good science article tends to pull others in its wake. It thus seems that, once enthusiasm is aroused among non-scientist colleagues who are thought to represent the great non-scientific public outside, the lines of communication are cleared, at least for the time being.

No reasonably objective observer can seriously argue with the view that a great deal of science writing is poor. The problem is not so much one of trivialization; one does not expect a mini-scientific journal hidden within the bowers of a daily newspaper, especially when it is the paper's explicit policy not to cater specifically for those interested in science. The average science correspondent tends to concentrate on his own idea of science to the exclusion of all else, and often fails to realize that what to him, and probably to others, is of great importance is of little value to his colleagues and superiors in terms of their perceived roles. The average science item deals with a small portion of a greater reality, and in terms of newspaper values it gets the treatment it deserves: an attitude also reflected in the feelings of many scientists who apply critical judgement to the social value of their work.

But even if every encouragement were given to bigger, better, and more detailed science stories, it is questionable whether many of the science-writing fraternity would be able to cope. In their more sombre moods this is clearly realized by the science correspondents themselves,[32] but unfortunately the enunciation of the problem brings it no nearer to solution. The essence of the difficulty that the human side of science writing has to overcome is not that there are no outstanding individuals in the trade or profession, but rather that because of the poor career prospects few of them choose to stay in it, if they have any choice. Not only is science writing paid at very moderate rates compared with normal middle-class earnings, but its career prospects are nil. The starry-eyed novice discovers after a time that, however brilliant his stories and however warmly he is commended by the editor, he will always be a science writer, and—should he exhibit the size of ambition perfectly accepted in a manager in industry or a barrister at court—he will have to move out from his chosen territory to become what is laughingly described as a 'newspaper executive'. Under these circumstances it is hardly surprising that many of the best brains or the strongest characters tend to drift out of science writing. How such a state of affairs can be cured, or whether indeed it should be, given that there must of necessity always be more privates than generals, is impossible to argue. It seems, however, to be self-evident that, all the time the science correspondent is regarded as a second-rater, in a sense he will internalize the image received, although—again not surprisingly—not altogether with good grace.

That matters can be arranged in a more equitable and civilized manner can be seen from the example of at least one paper, *The Times*, where on major issues the voice of science is expected to be heard, at least as part of the multifarious contributions to an over-all policy. *The Times* also claims a fair degree of collaboration between specialists and general staff, including the numerous foreign correspondents. This more agreeable state of affairs probably results from the ability of individuals to agree on a working compromise which, *pari passu*, recognizes that the paper is mainly about political and social questions and that the specialists—economic, political, scientific—can play only an essentially peripheral role. In the last analysis, a science correspondent works on a newspaper if he implicitly agrees to the limitations of his work there; he is there because his time—although not his views—is bought, and, despite the sneers of the left, that is precisely what happens to all but an insignificant minority in any occupation. Newspapers are about instant or nearly instant information, not education, and the context necessary for any continuity is missing. However, having established a working atmosphere of mutual respect, if not social equality, it is possible to exert some influence on the paper's policy on an *ad hoc* basis—which is about the best that may be hoped for under present constraints.

As currently constituted, newspaper publishing is innately not inimical to the presentation of scientific news and—more importantly—of the background ideas that give science its cohesive strength. Rather, possibly of necessity, newspapers do not speak the same language as even a diluted and trivialized science would require. There are a number of reasons for this, mainly historical and self-delusive, but paradoxically one of the most trivial exerts the strongest pressure. This is the general obsession among newspaper people with techniques of presentation, as if they in any sense delineated the very essence of a publication. It is true that over the years newspapers have changed and a paper of today is vastly different in appearance from the same paper ten years ago. But the most piquant point about design and presentational features is the faith placed in them. For example, the editor of a vastly successful popular paper will rhapsodize at length about the need to present every story in no more than 12 paragraphs, on the argument that their readers would not be able to read any more.

Now the intellectually arrogant may reply that this is absolute nonsense, and that the magic number 12 is a figment of the editorial imagination which any rational argument could tear to pieces. Not a bit of it. At the other end of the scale, and in its terms an equally successful publication is infinitely proud of its own stylistic trick. This is the 'delayed drop', defined as the ability to 'delay the news point as late as the third paragraph'. To the outsider, to one with even the slightest acquaintance with the infinite riches of the English—let alone the European—literary tradition, these arguments are incredible. But one must again take into account the *modus operandi* of all newspapers and the need to have absolutely firm guidelines which provide rough and ready criteria usable under conditions of stress and urgency when experience, that is reliance on a set of internal standards, is the only way to make decisions with any hope of consistency.

Newspapers must, every day, rush in to fill a vacuum, and within that major limitation the structuring of each publication's style into rigid categories is understandable if not laudable. Allied to the need for technical regimentation, there is also one for intellectual simplicities. Talking with Fleet Street, one cannot but be impressed how much better newspapers are organized for reacting to the tangible, the physical event. A catastrophe, a war, a political *démarche*, calls for vast feats of organization and communication vanquishing all difficulties. A similar event on the intellectual plane—a change of emphasis, a new thought construct— is a different species, and the well-known and smoothly operated mechanisms are useless. Thus every story has to have a peg: a real, understandable, exciting, or shocking event that can serve as the text for the lesson that will support the intellectual ballast. But science is not, fortunately perhaps, about bleeding hearts and falling empires; although its effects may be equally fatal, its first announcement is almost invariably *sotto voce*. One must argue that it is this particular innate quality of science, this divorce from the obvious edge of human experience, that must make it extremely difficult to present it in terms of a newspaper, quite irrespective of the intellectual level of the paper itself. This statement can be argued in the absence of any value-judgements on the quality of science writing, of the appreciation of editors and others, of the importance of science and its role in editorial policy or desirability as a current awareness method; it is simply that science, by its very nature, is not such as can be dealt with to any degree of satisfaction by the methods of newspaper production, at least as practised in the English-speaking countries.

Despite its chillingly satisfactory nature, this argument can be stood on its head by asking, Is not science part of the culture of the world we live in, and are not newspapers the true messengers from the centres of the system? The invariable answer is in the affirmative. Indeed, science is part of our culture and we should treat it as such; but what should we do? Thus one of the criticisms made of the *Times* science reports is that by putting science into a small ghetto of its own they contribute to its isolation, and nobody reads them. What about the other specialists—the theatre critics, the arts critic—whose reports from their particular manifestations are often as technical as those of the scientists but whose contributions are regarded as proper material for presentation day in and day out? Indeed, but they are tangible events and (unsaid but nevertheless firmly implied in the background) they are, if not gentlemanly activities, at least understandable (or should be) by the average arts graduate from whose numbers most of a newspaper's top executives originate. We are therefore presented with a contradiction: the idea that science is a part of modern culture is warmly embraced, and great regret is expressed for inability to do much about it; at the same time there is a consensus among newspapers from the most trivially popular to the most intellectually penetrating—or perhaps only pretentious—that science is an area of human activity best left to its own devices if it cannot be used as a filler for another and more immediate subject. In this attitude the newspapers are greatly encouraged by the almost total absence of readers' reactions to matters scientific as expressed in their letters to the editor. Exceptions are sundry instant-glamour subjects; the environment and insults against it was at one time a happy hunting ground for both the concerned and the crank; energy and resource crises, although primarily political and economic, allow the intrusion of scientific and technological facts and speculations; the wisdom of new hardware is occasionally questioned; and puerile letters appear drawing attention to the number of Nobel and other prizes won by particular classes of scientists. But all this does not add up to a concerned and interested readership, and because the first task for survival is the maintenance of circulation, newspapers have little enough motive to change their present stance.

Science in the press

How far does the present uneasy fellowship between scientific communications and newspapers matter? Does what newspapers say make the slightest difference to the course of events? Newspaper editors themselves seem highly ambiguous on this score. One point of view holds that newspapers have little power; they may be able to increase an already existing trend, but are certainly incapable of starting one. On the other hand, the role of newspapers as providers of information must be counted as a catalytic process, especially if the information would not normally be available unless the newspapers were determined to obtain it; there are several recent examples. However, in matters scientific the newspaper cannot set itself up as a judge and must have an essentially passive role, since the validation of the importance of an item cannot fall within a paper's ambit. An extension of this argument would contend that newspapers are not about reality at all, but a certain shadow of reality that is built up on a consensus of *ad hoc* decisions and imitations. On any one day virtually all newspapers, irrespective of intellectual or social level, will place all incoming material in roughly the same order of importance.

One could therefore argue that for the popularization of science newspapers present an unimportant side-issue. The papers themselves argue that their role is not educational and their mechanisms of production are not organized for the continuity of presentation that is essential in all educational processes. Warming to this topic, they will assert that newspapers are not about reality at all, but about something much more insubstantial: a relationship of dreams and perceived opinions, a well-oiled machine operating increasingly in a void. It is a neat picture, oversimplified to the extent that it cannot but lose touch with reality itself, since even a superficial examination of the style and content of newspapers over the last ten years shows appreciable changes mirroring, even if not leading, that indefinable entity, public opinion. But how far a shadow of a shadow can contribute in any way to the general understanding of science is a question of quite a different order of magnitude to which ready answers are certainly not available. The effective part of a newspaper's operations are limited in any case; the greatest triumph of a science correspondent's story is a question in the House which more often than not will be turned away by a judicious banality—fairly small beer, one would have thought, for the amount of effort it had required. Yet, as one editor has remarked, the curiosity of the public remains, and newspapers exist to satisfy it. From their point of view it is ultimately immaterial what stories fill the gap, and science, together with any other, will do. Thus once again the onus is thrown back on to the scientist and his method of communicating about his work.

Magazines

If newspapers do not succeed in making science part of the common weal, perhaps the innumerable magazines and journals could be brought into play. Unfortunately the number of general science magazines in Britain has decreased to one, hardly a complimentary reflection on our present state of scientific interest. *New Scientist*, which *en route* has digested *Discovery* and *Science Journal*, is the lone representative in Britain of a genre that is still somewhat more populous in the United States, and, despite smaller national populations, appears to lead a healthier existence in Europe. From all subsequent analysis the U.S.S.R. is exempt, since the leaden boredom of their didactic publications is surpassed only by the gigantism of their circulations.

When the average scientist thinks about a general science magazine, he is apt to think of *Nature*, or perhaps of *Science*, its American equivalent. He thus demonstrates that he has no idea of what popular communications are about, for both these publications exist primarily because they offer rapid publication for short research articles, heavily biased towards the life sciences. Both also feature items about quasi-political matters with variable degrees of success. In general, *Science*, which operates in the much more open—or indiscreet—American environment, can be relied on to report the political or organizational origins and consequences of scientific matters. *Nature* has achieved a successful monopoly of the *ex cathedra* statement, as befits a self-appointed institution of the realm, neatly falling between the interest of scientists in science and of non-scientists in the effects thereof. It is noticeable that, although both *Science* and *Nature* are read (if not regularly) by science correspondents, the source material used resides in the scientific articles, not in their opinion columns. But both publications are virtually unknown to the general public.

IPC's *New Scientist* is successful, in a moderate way, signified by a circulation gently undulating between 60 000 and 70 000 which, together with its advertising revenue, is sufficient, at least for the time being, to preserve it from the fate which had befallen its erstwhile rivals. The difficulty with *New Scientist*, as indeed with all other magazines in this general field, is that in no sense are they popular. The readership is overwhelmingly among scientists or those who have already some interest in science or its effects; thus schoolteachers form a valuable substrate. It would accordingly be difficult to enjoy more than a fraction of articles appearing without at least a modicum of scientific knowledge, although, as a yet-unpublished survey has pointed out,[33] even scientists

Science in the press

have no objections to topping up their ready supply of quotable quotes and opinions on matters they know nothing about.

Scientists read these magazines 'partly out of intellectual curiosity, partly for conversational value and partly to maintain social contact with other scientists . . . . Most . . . see library or circulated copies but do not see every issue. Usually they scan the pages in sequence and typically look at between a quarter and three-quarters of the articles plus the news and comment, book reviews and classified advertisements.'[33]

What is true of *New Scientist* is even more so of the plethora of magazines which the National Union of Journalists calls with somewhat justifiable realism 'trade and technical'. In publishing terms, they range from those presented by learned societies and institutions to commercial ventures, and their generally distinguishing feature is that they take as their domain a fairly small and identifiable range of topics. Among the society publications there are some with circulations any commercial publisher would be proud of, and from a presentational point of view most of them have now caught up if not with the 1970s at least with the 1950s. Their contents have also improved a great deal over the past ten years. It is now accepted that even a society publication need not concentrate exclusively on the trivia of the local sections or the formalities of administration. The better ones have come to view themselves as meeting places between specialized sections of the scientific community where the social, economic, and political factors affecting science may be discussed, together with general developments in a particular science. Society magazines have therefore come a long way in assisting the social, concerned scientist, but the educational process is directed towards the specialist, rather than towards the general public. The fact remains that, because they are society publications and are therefore part of the deal— and myth—of belonging to a self-defined community, it is difficult to measure in quantitative terms the number of real readers, in contrast to the wastepaper-basket fillers. In two cases it has been shown reasonably conclusively that 95 per cent of members of particular societies do scan through their published offerings, but even in the best case the domain is not one of acquainting non-scientists with science, but rather of acquainting scientists with events in the world that struggles on outside the laboratory.[33]

Among commercial journals, the presentation tends to be better and the intellectual content worse—not surprisingly, since they have to live on their advertising revenue, accurately keyed to their editorial policies and nature of subscribers. Some ten years ago the so-called controlled-circulation magazines took this theorem to its logical conclusion by

Science in the press

sending the publications free to specific groups who in their opinion presented the best and most homogenous targets for advertising. The rise in postage rates, increases in all other costs, and a slow realization that a gift horse is rarely looked in the mouth has led to the abatement of this trend, but there are still a number of these journals in existence. Because of commercial pressures, although the best of the commercial specialist trade and technical magazines can be very good indeed, the bad are quite outstandingly so and the editorial matter appears as filler between the advertisements. Their specifically limited circulation ensures that these publications are expressly for the specialist, and their effect on the general knowledge or appreciation of science, or indeed of anything else, is nil.

One need not be over-pessimistic in concluding that the present state of newspaper and periodical publishing is hardly the best way of making a non-scientific public aware of science. The faults are multiple, but nearly all of them arise from a series of historical accidents and the need, as far as the publications are concerned, to survive in a commercial environment. Despite the myth, the crusading publication is a rare bird indeed, and even then the crusades are carefully chosen so that moral uplift should harmoniously match commercial gain. Neither is it very profound or sensible to assert, as some critics do, that the system, having been found wanting, should be destroyed, for no greatly improved one has been found to take its place. The only present alternative is a massive subsidy from public funds, a mechanism that is operated successfully in socialist countries, where the role of the Press is viewed in a very different way. But if we still desire a multiplicity of choice and opinion, then the commercial Press is still the only viable alternative. The science communication enthusiast must therefore come to terms with it, *pace* trivialization, distortion, and rushed deadlines.

4
Hard and soft science

Many of the difficulties in the path of communicating scientific ideas to those who are not scientists are economic and organizational. There is nothing specifically scientific about them. They relate to organizational structures, to making decisions in a void, and to ill-quantified feelings among those placed in a position of influence. One newspaperman expressed a profoundly pessimistic—or perhaps only realistic—thought about the organization of all publishing when he suggested that the ability to achieve high position includes 5 per cent talent and 95 per cent the knack of being in the right place at the right time. Thus the woes of science are identical to those of all specialist interest areas; snooker enthusiasts could criticize publishers and public in exactly the same terms.

Beyond idiosyncratic aversion and delusions of populist grandeur there remains a domain of genuine difficulty in communications about science. The difficulty does not, in this instance, relate to arguments about the existence of one or more cultures, or the relationship between average educational attainment and the intellectual level of contemporary science. It relates to the emotive aspects of science.

It would be seriously misleading to suggest that large areas of science are not utterly boring. On hearing some of the *eminences grises* of science, especially those who have not been in a laboratory for the past ten years except to cut a ceremonial tape, expounding on its interest, thrill, and intellectual challenge, one would imagine that all practitioners of science are perpetually on the brink of some spiritual orgasm, occasionally shouting Eureka to relieve the unbearable tension.

The truth is that large areas of science are tedious. The tedium of learning a discipline has been generally accepted in the past, and, although some theories of current education suggest otherwise, it is still accepted in the humanities, in the arts, and in most of science.[34] In the same way that learning to play the piano inevitably includes decades of Czerny, learning to be a scientist must of necessity involve the acquisition of intellectual and manual skills through practice rather than through intuitive exploration. What is not, however, sufficiently appreciated is that not only the learning but also the practice of the sciences requires a great deal of patience, perseverence, and the ability not to fall asleep when guarding the umpteenth repetition of an experiment. At the edge of science where discoveries are made there is indeed intellectual excitement, but more often than not it is accompanied by lengthy and repititious experimental programmes to make sure that the discovery is correct, and not an experimental error writ large. Neither does it matter if a great deal of work which previously was carried out by the individual is now performed by machines; only a very inexperienced or foolish scientist would place complete faith in his machine to the exclusion of personal vigilance.

Added to the necessarily tedious practical work, there is also a need for the often-boring work of erecting intellectual scaffolding. No matter how many computers there are at hand, ultimately it is the scientist who has to pore over results to correlate and assess, to find errors and make the necessary adjustments. No wonder that a large number of scientists lose their appetite for the humdrum, if necessary, aspects of their discipline as soon as decently possible and relegate these tasks to the research assistant or the junior staff.[35] The folklore of scientific research is full of professors who see their research students once a year and whose absence from the laboratory is an accepted tradition.

Even making allowance for human preferences in avoiding the boring and the tedious, it would be untrue to suggest that scientists are enthusiastic about all aspects of even their own fields. The decline and renaissance of particular disciplines and sub-disciplines is a matter of record, and, although many instances can be explained by the presence or absence of the necessary theory or appropriate apparatus, some are undeniably caused by pure fashion. Scientists have therefore few grounds for complaint if the general public finds large areas of science boring or trivial. It might be better to consider a point of departure which stipulates that significant areas and aspects of science are not fit for public consumption. The effort to communicate should be reserved for carefully chosen aspects of science where the possibilities of success appear at least reasonable. Such domains are generally represented by technology rather than by science, since its physical impact on the actual or potential environment is rather easier to appreciate than the more fundamental if more tenuous results of science. But the relationships between science and technology are complex, and only a few people still argue that technology is a direct consequence of science.[36]

Even if the topics within science suitable for general interest are carefully selected, the difficulties are only beginning. One popular myth in the reporting of science goes along these lines: 'Scientific papers are written in a formalized style that is difficult to read. Their formalism obscures the most important facts of scientific research, the succession of failures and successes, and imposes an air of logic on scientific development when it would be more illuminating to consider the intuitive factors in scientific practice.' Closely argued articles are written on the topic, the editors of scientific journals are berated for allowing the stylistic monstrosities of their euthors to roam the countryside, and indeed an attempt has been made to produce a vernacular version of an erudite physics paper.[37]

It is good to know that those who talk about science have taken Karl Popper to heart. But the communications within science itself, as exemplified by the research paper published in the scientific (primary) literature, represent a complex of motives, aspirations, and pure logic. Its role within the scientific community has been written about and investigated.[37a,b] Whatever its roles for the professional scientist, it is clearly not designed as a vehicle of information for those outside the scientist's particular specialization. Although science journalists occasionally use original papers as primary material, their importance in more general communications is small. An obvious proof of this statement is the proliferation of scientific series on the lines of *Advances in . . ., Progress in . . .*, which have been spawned in recent years to the greater satisfaction of authors, libraries, and publishing houses alike. Thus, whenever one hears the lament that the style and content of scientific papers is a prime cause of difficulties in transmitting science to the uninitiated, one should immediately be on one's guard; the protagonist either has not read a scientific paper for a long while, or amid his other preoccupations he has totally forgotten its original and still valid purpose.

Rather than rely on the primary literature of science, the communicator will more often look through the various review publications which have already placed the new material in context. More important, however, is the scientist himself, who, with a little coaxing and a reasoned show of interest, can usually be persuaded to guide the hobbling enquirer in the right directions. Having obtained his raw material, the interpreter then has to make some valid use of it, and here we come to the central problem.

The problem is related to the availability of signposts along the road, which in turn is a function of the particular science being considered. One can talk of an obvious example of this argument if we disregard science and relate it, for example, to the process whereby a Chaucerian piece be explained to a reader of the *Daily Mirror*. Although the language of Chaucer is no less unfamiliar to our stipulated reader than the latest phraseology of solid-state physics, the matter described does relate intimately to everyday human experience if we assume that the processes of procreation, birth, maturation, and death have changed but little during the time-span we are interested in.

Conversely, Boolean algebra, although it may be the most powerful tool of logic brought to bear on the problems of science, unmistakably lacks an obvious and readily identifiable handle for holding it up to public admiration. Between these two extremes the sciences cower, waiting for the interpreter who can reduce the severe formalism of mathematically based topics to the familiar terms of everyday experience, without at the same time introducing inaccuracies which would render any scientific aspect completely misleading. In a sense, therefore, one might consider the ease of conveying scientific information as a function of the nearness of the domain to common human experience, dealing with events which, at first or second hand, have built a tangible, emotional nexus to the recipient public.

Needless to say, on this argument medicine and the life sciences are high in the league tables. Most developed nations consist of innumerable hypochondriacs, crunching hypnotics on going to bed, pep pills on getting up, and tranquillizers to help endure the frustrations of public transport and cantankerous managers. It is not surprising that any medical or paramedical information obtains a ready readership. Although its entertainment value might not compete with confidences exchanged in the surgery, it is a close second. Similar arguments apply to the life sciences, especially to those that adopt the whole-animal approach of classical biology and zoology and those that appeal to the discoverer and cataloguer in all of us.

Two very different books may be considered, each of which enjoyed a very great success on the British market, but for very different reasons. W. Keble Martin's *Concise British flora*[38] was turned down by innumerable publishing houses, who perceived the costs but not the potential profit, until, rescued by a sudden flash of insight, it was published only to become a bestseller. By hindsight, it is obvious why it should have been so. Despite the difficulties of jargon and definitions, which in any

Hard and soft science

case need not be taken too seriously by the amateur, it is a book that invites the participation of the reader by asking to compare experience, by sharing a common interest. As the age of increased leisure floods in, Victorian activities become magnified and mass-produced: the new participatory age for amateur botanists, railway enthusiasts, anglers, and Sunday painters invites the hordes of enterprising publishers.

In this context it is interesting to consider that Keble Martin's *Flora* neatly sidestepped all the common wisdom of popular science publishing, which perhaps accounts for its initial reception. In material terms, its proliferation of coloured illustrations meant high production costs; thus economically it could be considered viable only if very large numbers of the product were sold. More importantly, it has an explicit and unashamedly didactic tone: it is an uncompromising package of information, uncluttered by considerations of social, vocational, or political undertones, or by a need to flatter the reader psychologically or intellectually. Yet these very characteristics which made it look so old-fashioned and unsaleable are its strengths. Because it is not utilitarian, because it does not shout an ideological imperative, it allows the reader to enjoy himself without feelings of guilt and, more importantly, to associate the pleasure of the book with the pleasure of his chosen relaxation. The sight of National Trust land in spring and summer full of unskilled parents attempting to identify flowers for their uncomprehending children may not signify the rebirth of botanical discovery, but at least it constitutes a healthier and more self-instructive occupation than passive technological or commercial mass entertainment. Although superficially this type of publication appears to demand intellectual exertion (for example, becoming familiar with the vocabulary of botany), in reality this is not so. The book may be enjoyed by an intelligent eight-year-old. He may or may not wish to make the effort of learning the technical terminology of botany; his total experience will hardly be affected. Neither does he have to ponder the underlying questions of ecology, genetics, plant physiology, and geobotany which would raise his knowledge from a hobby to a discipline; all that is unnecessary.

It is dangerous to attempt to draw any conclusion from the fate of one statistically unrepresentative book, but in a qualitative way one can suggest that at least some of the factors needed for a successful communication of some aspect of science to the general public may be found among: an approach tied to individual leisure or enjoyment, an avoidance of deep issues, a topic which is easily within the intellectual compass of most potential readers, and finally—but most importantly—

a pleasing style and presentation which reaches to the very fundamentals of book production, the need for the book to be a pleasure to handle and to read.

Another successful example of scientific communications in this domain is Desmond Morris's *The naked ape*.[39] It is a much more complex and sophisticated book than an encyclopaedia of flowers, yet in publishing terms it has been a comparable success. For this argument it is irrelevant if the facts presented or their interpretations are correct or not; and there does appear to be evidence that serious errors of fact and interpretation exist. The book belongs to the school called variously ratomorphic or chimpanzoid, which attempts to explain, or at least illuminate, human behaviour by reference to animals. In *The naked ape* the most memorable examples relate to sexual and aggressive behaviour: essentially the same material that provides the basic material for authors as different as Zola and Grace Metalios. But whereas literature does— or at least is expected to—provide insight into what Hobbes called the human condition, the ratomorphs, under the label of explanation, provide a philosophical justification. This argument is—and by its nature must be—contentious. Neither is it wholly fair to suggest that the intellectual voyeur is more easily accommodated to the contemplation of sexual activity if it is clad in the academic respectability of scholarship.

All this matters little, for the naked ape, like the encyclopaedic flower, suggests that the easiest approach to the popularization of science lies along the pleasure principle rather than the intellectual challenge. The pleasure may be as varied as the conditions of the reading public or sections thereof; it might be the simple smile of childlike recognition as much as the application of a soothing balm to the guilt-ridden self-judged intellectual. In both cases the reader emerges, if not better learned, at least reassured. The world may not have attained the heights of Utopia, but at least it is unquestionably still there, and it is hardly our fault, bearing the burden of genetic guilt, that it is in the mess in which we find it. Thus, in common with the Catholic church, Marxism, and a few other creeds and ideologies, science does not only ascertain the miracles of nature; it also provides a ready absolution for our own shortcomings. The Blood of the Lamb is effortlessly replaced by the adrenalin of the ape.

The underlying strength of these approaches, of which perhaps we can consider these two books as extreme or limiting cases, is that they rely on a degree of familiarity with both the material and the conceptual framework. This characterizes virtually all popular science writing within

Hard and soft science

the life sciences. However limited our knowledge, we have at least a superficial acquaintance with the chief natural phyla of the world, and we have a rough and probably inaccurate idea of their characteristics. More important, because these sciences are only now emerging from the cataloguing stage, the conceptual framework has not yet attained the true difficulties of the physical sciences. The majority of the life sciences, including medicine, are far more part of our general cultural pattern than the physical ones, and therefore the process of translation assumes a much less important role. Since the majority of underlying concepts are already common, it is sufficient to increase awareness by examples and description and by drawing parallels; in short by piling up facts on existing foundations.

Such an approach works badly in the physical sciences. For a start, the emotional nexus is missing. One cannot speak of small furry atoms, the molecules of the wide open beauty spots, or the refractive index of the guilty human ego, and cannot therefore draw even an answering sigh from the factually overwhelmed reader. Neither is it clear how the dictum about science being no more than organized common sense could be interpreted in explaining the mathematical intricacies of subnuclear particles or the hidden beauties of steroid substitution. No amount of fine writing, intellectual fireworks, or the stretching of metaphors will make Heisenberg's uncertainty principle or the idea of kinetic equilibria quite as easy to swallow as science journalists would like. This is partly because of factual strangeness and partly because concepts derived from common sense no longer apply in these topics.

It has indeed been argued that the difficulty of the physical sciences not only frightens off the non-scientist but also makes him feel guilty and perplexed by implying intellectual feebleness.[40,41] This in turn renders him inimical to science. One can suggest that the behaviour of non-scientists can equally well be explained without the stipulation of complex psychological mechanisms engendered by the physical sciences. It would be simpler to believe that the non-scientist is simply bored by the whole business and wishes to have nothing to do with it, arguing that those whose job is to deal with the problems of science are sufficiently well paid already for taking on the layman's burden.

What should the science journalist's response be? On the one hand, he is reduced to suggesting explanations that are scientifically wildly oversimplified and misleading and, on the other, to despairing of the whole sorry mess and invoking the gimmickry of technology. The interpretative process in scientific terms is very rare indeed.

But, it will be asked, why not illuminate the advances of science in terms of well-known everyday, artefacts which form an ever-present part of contemporary life? At one time virtually every popularizing book or article on modern chemistry would start with a paean of praise for plastics and ask the reader in an initial sentence to look around the home and note how much of its comforts were due to the advances of the plastics industry. This is very simplistic thinking, for the reader's reaction tends to be: So what? Instead of immersing himself in the intricacies of thermoplastics and thermosets, he most probably turns to the nearest soap opera on television. The error in this argument is that it fails to acknowledge that no emotional nexus exists. The presence of every-day, well-known consequences of science is thus a necessary but not sufficient condition for interesting those whose inclination, by our own definition, is not scientific. It may be a fault of our cultural milieu, of our inadequate teachers, or of the insensitivity of those who form public taste, but the fact remains that most of us take the miracles of applied science for granted and do not consider them either beautiful or exciting, but merely rather irksome necessities which provide us with creature comforts to which we have become accustomed. On the other hand, it is possible that most of the miracles are neither beautiful nor exciting.

Whatever the moral or ethical point of view, the brute fact remains that outside the socialist countries only company chairmen rejoice at the sight of a larger and larger dehydrogenation plant—or at least rejoice sufficiently to make their happiness a matter for public comment. Not unexpectedly, the general public's attitude is largely negative, which causes a great deal of irrational heartsearching among those engaged in the unpopular sciences, who feel a distinct lack of public appreciation of their work. They should not worry. The lack of public concern is genuine, and may it long remain so, for if a camel is a horse designed by a committee, a new chemistry developed by public request would be a horrendous monster indeed.

Science journalists are always looking for new openings for physical science stories. A number have presented themselves and can be adapted to suit the prevailing climate of opinion. One approach which works well enough in the United States and the U.S.S.R. but less effectively in Britain is to refer to national pride. There appears to be some assumption that scientific and technological advances are in some way related to the greatness of a nation—whatever 'greatness' may mean under these conditions—and thus an opportunity missed is a derogation of national responsibility. This argument was used brilliantly during the space development in the United States, and somewhat more subtly in Britain in the post-war years when the future of atomic energy was being planned. But since the disastrous Labour references to the white heat of technological revolution, the hotly contested Conservative lunges toward technological masterworks, and a realization that big science is getting rather too large for most national boots, the nationalistic arguments to make science palatable have lost some of their force.

Their place has been taken by two contrasting topics: economics and the environment—the latter word being used in its contemporary and wrong meaning. There is still room for further development here, for it is incontestable that science has both economic and environmental consequences, and knowledge of one may thus shed light on the other. Neither does this argument suffer when it is occasionally pointed out that in operational terms the friends of the environment, according to their own definitions, are directly in opposition to the friends of economics, since the creation of a large power station may be good economics but is almost certainly bad environmental conservation. Whatever may be the rights or limitations of either argument, it should perhaps be pointed out that in countries where the meaning of hunger is still a general, ingrained knowledge, the ecological argument comes a very bad second behind the economic one.

Disasters sell newspapers. It would be quite wrong to assume that the same considerations do not supply, even if at a more sophisticated level, to news about science. The disaster may be tangible or potentially so: the possibility of a nuclear reactor getting out of control or a harmful chemical being emitted into the atmosphere.[43] But beyond the voyeur of disaster, the bored observer of the antics of government, beyond even the intellectually slighted *homme moyen sensuel,* there is a shadow of fear. Fear of the unknown, of being manipulated, of forces liberated by Them whose direction we cannot control, of an unbearable rate of change. It must be argued that this element of fear provides the engine both for a great deal of anti-science opinion and also for a desire simply to disregard the whole present-day apparatus of science and technology. To most scientists, however socially concerned they may be, such an attitude is simply beyond understanding.[44] The science journalist is caught between opposing forces. To make his story interesting, he must establish an emotional contact with his readers, but this will unavoidably bring him in conflict with the scientist whose goodwill he must preserve. At the same time, his own inclination is to present science as a rational human activity, beneficial despite instances of misuse. No wonder that the balancing act is often less than successful.

In actual fact, the best science writing is committed to a point of view that includes not only science but also society, in which writing about science forms only a part of the whole.[45] Such writing succeeds in establishing contact with the public through emotional sincerity rather than semantic accuracy expressed in pedantic lists of facts and definitions. In the case of especially the physical sciences this is vital. Science must be regarded as exemplary material, for the arbiter of history is man and not the other way round.

Exceptions to these statements are a very few successful attempts which, through the excellence either of the writing or of the visual presentation, arrive at a view of science which becomes exciting through its medium of transmission. Subjects such as astronomy, geology, and— for some—mathematics and physics can thus suddenly leap off the pages of the book or the screen of the television set and become experiences of pure pleasure. But it needs a Jeans, a Haldane, or arguably a Bronowski to achieve such a result, and even then it cannot be produced every time the practitioner so desires. The information content of such presentations is usually low; those of the audience who knew little about plate tectonics or the fate of the galaxy will not continue their existence with an assured grip of these topics. But this is irrelevant. In the ultimate, it matters little if the recipient of the science communicator's art remains no more informed than the geneticist's naïve worm. It is unlikely that he will be asked to recite the universal truths of Pauli's exclusion principle, or, if he is a political leader, that a knowledge of Heisenberg's ideas on the uncertainties of measurement would be a great help to him the next time the oil lobby becomes too energetic. All one can hope for is that these attempts at translating the ideas of science into the language of everyday concerns will illuminate some of the underlying ideas, dreams, and limitations of science. Nobody requires a poet to become a neurophysicist, although both are dealing with different aspects of the same subject, but it is undeniable that a greater cultural homogeneity could help towards a reasoned solution of some of our contemporary ills.

5

Science and television

Compared with newspaper coverage of science, television is in a different league of popularity. The national daily in Britain with the largest circulation is the *Daily Mirror* with a sale of somewhat over 4 million; but it is rather doubtful if anything like that number of readers actually makes its way through those 12 paragraphs' worth of science stories. The average BBC *Horizon* programme has an audience of 2·5 million viewers; and that figure arises from the BBC viewer research programme, which relates to actual sets in use—in distinction to commercial television's JINCTAR, which relies on a sample.

When pressed about science coverage, newspaper editors will point out that science and education are properly in television's domain, since the moving finger explains so much better then the printed word. There appears to be an internal contradiction here. Since television is so much more popular than the newspapers, and the latter shy away from science because it is supposed to be above the heads of most of their readers, how is it that television can demonstrate week by week that non–scientific viewers will devote their time to the mysteries of the DNA or the behaviour of fish? Perhaps the answer lies not in intellectual but in visual content—a point which will be explored further, and which is easy and uncontradictable. But as a starting hypothesis one should also suggest that greater popularity is perhaps due precisely to the fact that (at least at its best) television does not concern itself with popularity but rather with the professional standards of its presenters. Despite any amount of self-justification in managerial terms, the final criterion of a programme's success is in the self-regard of its creators.

'It is generally agreed.' In a research paper this phrase can be roughly translated: 'There does not seem to be much evidence for it, but nobody I talked to during coffee could raise any objections.' By a similar methodology, one can assert that, of the British television programmes dealing

Science and television

with science, *Horizon* has been generally regarded as the most consistently competent and enjoyable. In support of this statement one can quote the large number of sales overseas, the generally high viewing figures, and the air of unhysterical professionalism the programmes show. I propose to use it therefore as a yardstick for comparison with others, and to inquire into some of the reasons for its success.

For purposes of comparison, *Horizon* can be regarded as a magazine operation rather than as a newspaper; the number of people involved is small enough for all of them to know each other, and also for the kind of general consensus to emerge that is expected when professionals are working together. This consensus does not require anything as crude as a written-down set of policy rules, but it ensures a rough homogeneity of thinking. There are a number of other advantages also. Although it is claimed that no policy as such exists, the unit does know in advance how many programmes it will have to generate in a given time. (New programmes are required for roughly nine months in a year, the summer being taken up with repeats and miscellaneous programmes.) Such a limitation is a necessary discipline, since there must be some understanding of rough proportions between disciplines that ought to be kept. More important perhaps, the obsession with news that haunts the newspapers and the magazines is absent, perhaps because of a true understanding of the irrelevance of this concept in popular science, more probably because the translation into a new medium assures news values *per se*. *Horizon* was given considerable technical scope by the decision some years ago that it should go on the air weekly and that it should be shot on location rather than in the studio.

How does it all start? Surprisingly, this is not a question that can be answered with any degree of assurance. Some ideas are snatched from the air; in an ongoing operation, as one programme is being made ideas and remarks are tossed about which may suddenly land in some producer's collecting-jar. There is also much more editorial planning than on the newspapers; *Horizon's* editor claims that he spends a considerable amount of his time going through scientific journals and magazines to find out what is going on. Curiously, at this stage there is very little contact with people who, one would have thought, could give a far more up-to-date account of new developments in particular sciences. But this time in the formation of a programme can be regarded as the induction stage when, using an old-fashioned phrase that they themselves would reject with scorn, the team waits for the muse to bring her magical gifts. In more mundane terms, they wait to see if anybody is getting sufficiently excited. If the excitement becomes infectious, a more-or-less objective rationale may be grafted on to the original idea. Since the technique itself is so subjective, it is difficult to find perceived criteria being used: statements about interest value being more important than news value are little more than *post hoc* justification. Nevertheless, the system appears to work, possibly because in this context an open mind, armed with sufficient money, time, and a professional knowledge of what the medium can or cannot do is not a bad prescription.

Once a subject is decided on, the actual making of the programme is carried out by a production team, usually a producer, a researcher, and a secretary. There are altogether ten or eleven production teams in existence, each having up to 14 weeks for a programme—a reasonable time for second thoughts to be incorporated. It is noticeable that within the production team structure central control is much tighter than on newspapers and on many magazines. Thus the idea has to go through a number of stages before complete acceptance, most of them revolving round a simplified participatory system. Altogether *Horizon* holds staff meetings of four different kinds: evening meetings every two months for informal discussions; general meetings; meetings for organization and production; and producers' and researchers' ideas meetings.

It could be argued that such an organization resembles a Chinese collective rather than a unit for producing works of education, entertainment, and, it is hoped, art, but the argument is misplaced. It has been one of the greatest tragedies of British newspaper and magazine production that those in management control have tended to regard the production of a page of print as being strictly comparable to that of a pound of sausages, despite the history of all publications which show only too clearly that quality is more dependent on the participants' desire to achieve it than on any other single factor. Unfortunately quality, especially in communications, is impossible to measure, certainly in the short term. The extra managerial effort required to give every person in a large organization a sense of belonging, a sense of being fully stretched, is difficult and in accountancy terms unjustifiable. Thus it is seldom seriously attempted, though it is easy to show that there is a close correlation between the morale of a particular team and the quality of its achievement, given the over-all limitations of a system. Following this argument, it must be contended that, consciously or otherwise, one of the chief factors in *Horizon*'s success has been the superficially cumbersome organization that allows everybody to have a say on ideas and techniques, however illusory such a participatory mechanism may be; for the editor still has the final word, and the necessary economic limitations are accepted.

The idea having passed through the sieve of contacts, conversations, enthusiasm, and meetings, it comes to the treatment stage, where the actual sequences are planned in reasonable detail. By this time, the best part of £1500 (1975 figures) has been spent, and thus, although an idea can still be turned down, in practice this very rarely happens. Already, before the treatment step, what is laughingly called research has been taking place and will have continued up to the completion of the programme. Precisely because it is television and thus able to play on human vanity, producers find no difficulty in interviewing virtually anybody they care to name. So much so that 'One can see the topic after a few days' phoning'. Such a statement could be an indication of utter triviality of mind or an incredibly rapid grasp of complex scientific matters; on the other hand, it may be an implied compliment to the scientists' often maligned ability to break down their subject into manageable portions. One must suggest that the last explanation probably comes nearest to the truth, for the scientist is dealing with a visual communication; he can point out bits of apparatus and experimental subjects; he is released from the residual constraints that the writing of formalized papers imposes on even his popular communications.

There is also a very important difference between TV science programmes and their equivalents on magazines and newspapers. Each production team is allowed up to 14 weeks for a programme, of which about a month is spent on research, three weeks on filming, and six weeks on editing, including the generation of graphic sequences where these are necessary. Comparing such a sequence to a newspaper article, which must react to the day's events and be ready by about six o'clock in the evening, or to a generally understaffed magazine, where if an article is less than perfect there is often little chance to change it completely in the time available, the television science journalist certainly has a more reasonable task. There is also a fair amount of money—possibly small beer by comparison with the film industry or the wilder shores of the entertainment business, but decidedly mouthwatering for anyone on newspapers or magazines. A science programme on *Horizon* will be budgeted at about £18 000 an hour (1975 figures), which compares favourably with the money spent on a run-of-the-mill serial, at about £25 000—30 000, and is up to a fifth less than the hourly cost of a full-scale drama or entertainment programme.

Science and television

Yet another factor working in favour of science on television is that the members of the public who are loath to write to their favourite daily paper on matters scientific are apparently more than willing to join in a visual argument. Up to a thousand letters may be received in the wake of a programme, although more usually it varies between twenty and thirty—which is still an enormous number compared to other media. There are also audience research processes for all television programmes, and, although one may disagree with the details of their findings, they are indubitably able to supply some indication of trends in audience response. Such a greatly improved channel of communication is important for the programme presenters; it may limit their scope, but it does remove the awful vacuum in which most magazine and newspaper work takes place. A professional, however creative he may be, works best when under the direct or indirect pressure of his patron—in this case the statistics of viewing figures.

One should note that audience research figures must relate to a programme already broadcast. It has been argued that an extension of such research may provide the producers with a better indication of the topics their potential viewers would like.[46] There is an underlying assumption in this argument that a coherent demand for certain types of programmes may emerge; but this is most unlikely. There are already a number of (generally polite) pressure groups in the science field trying to interest radio and television people in their disciplines: pressure groups based on learned, professional, and generally protagonist groups. Up to now their success rate has been minimal. Reid[47] has shown the impossibility of bringing the scientist too intimately into the development of a programme: production by consensus is an impossibility and in the last resort the responsibility of the producer cannot be shared. The producer's loyalty is to the programme, not to science, and his ideas, thought-processes, and value judgements are of necessity different from those of the scientist.

Enthusiasm for the job is helped along by the reasonable, if not outstanding, career prospects television offers its practitioners; by early middle age an experienced producer can expect to earn around £8000 a year (1975 figures), which is appreciably more than his equivalent on either magazines or newspapers. Creativity and competence may not be exactly measured by the amount of money they bring, but it would be false romanticism to suggest that reasonably good pay is not one of the best means of boosting the professional ego.

The general picture one gains from talking to those in television science journalism is a reasonably optimistic one, at least for the makers.

Significant criteria for measuring the effectiveness of communications are missing, and little effort is made to evolve them; the urgency of the daily task always takes precedence. Subjects are chosen on the basis of 'we know they are interesting', which a realistic translation would suggest means that the producer's wife agrees with the idea. Notions on social influence are still in their infancy and little articulated, although television can claim some successes; a rather trivial example concerns the malaise of pet-food manufacturers after a programme on whales.

What does seem clear is that, whether by good chance or a series of small planning steps, this area of the service has enabled its practitioners to have enough time and money to breathe, to spend some time in considering presentations in relation to each other and to have some ability to experiment and risk the occasional failure. Under these conditions, partly because of the nature of the people involved, and partly because a new medium is probably more open to the acceptance of varied ways of presentation, there seems to be a much better team atmosphere and much less anxiety in keeping the flag flying against pressures from other parts of the organization. Whether this state of affairs will persist in the long term it is hard to say.

The ultimate judge of the effort, invention, and decisions—good, bad, or indifferent—is the viewer. The chief attraction of television is its innate characteristic of movement; as A. V. Hill has said, if it wriggles, it is biology.[48] Thus the visual can, and quite often does, completely overshadow the factual content, and to those seriously interested in some aspect of science there can be nothing more irritating than the loving interplay between camera and subject which has little relation to the topic under discussion. This very quality of verisimilitude hides the greatest illusion, for the reality of the camera is not true reality, and in scientific programmes one is in effect questioning the skill of the editor in delineating the shadow of a shadow. No wonder that accusations of triviality are made of all television programmes; there can be little possibility of reflecting in them all the qualifications and ambiguities of science in the making.

Television, rather like the mediaeval church, has discovered that mass communication is possible through images which need not be accurate, provided that the general impression approximates to what is desired by the originators. It is hardly surprising then that the majority of modern states, rather like the England of Henry VIII, have taken steps to keep under state control the most powerful communication medium of the day. The visual aspect of television transmissions has a multitude of effects. It removes the need for formality that the written word requires, and (especially important for scientists, who always feel the shadow of the learned paper) it does not require the dignity of an *ex cathedra* statement. Thus, the spoken word or the taped image provides even the inexperienced science popularizer with an opportunity of conveying his views to a large viewing public, especially if his infelicities, inarticulateness, and a deplorable tendency to scratch his ears can be skilfully and invisibly edited away.

Yet the accusations of trivialization must stand, if only for the simple reason that an hour's programme can hardly summarize the accumulated wisdom of even a small fraction of a discipline. But even here television scores by the very evanescence of its changing images. Whereas an article in a newspaper or magazine can be read slowly and thought about, only the nimble-minded or the expert can follow a TV programme and simultaneously analyse it for sins of omission and commission. In this respect, television has successfully demonstrated that programmes can be presented on virtually any scientific and technical subject. If after the event the naive viewer's understanding is not much greater, at least some of the ideas have come across, or at least some spark of interest has been aroused.

One would have thought that the sciences, based on experimental work, would have offered admirable raw material for programmes such as *Horizon*, the visual interest illuminating conceptual elegance. But such an approach tends to be defined as an illustrated lecture, which is regarded as boring and old-fashioned, and instead a number of other techniques are used to focus the viewers' interest. Foremost among them is the human angle. This presents a difficulty since, although what scientists do is often of great importance and interest, the people themselves are not necessarily exciting and their appearances on the screen have therefore to be carefully rationed.

If one discusses the possibility of a science programme with broadcasting people, one of the first questions asked concerns the availability of lively and photogenic scientists to put the ideas across. Those who would persuade the broadcasting companies to present their most important ideas spend an unconscionable time scrabbling about their friends, acquaintances, and distant connections to discover the appropriate mix of voice, face, and transmittable vivacity which would enable the idea to be expressed on the screen.

Another approach could be termed the 'science spectacular', of which the BBC's Calder programmes,[49] and in particular the Bronowski series,[50] are good examples. Taking the latter as the logical quintessence of all such approaches, one remembers beautiful photography, the narrator's particular point of view, and very little science. Nor is this unreasonable. A series of programmes attempting to cover the total development of scientific thought, or even a complete discipline, as was the case in some of the Calder programmes, must be selective. But selectivity depends not only on the innate importance of a scientific idea, but also on the personal predilections of the writer or presenter, the availability of sources of information, and, inevitably, on the photogenic nature of the material. In terms of entertainment such programmes can be superbly successful; in terms of transmitting scientific information their outcome is more questionable, if the yardstick of success is the viewer's greater awareness of either the facts or ideas of science. It can be argued that lavish presentation is ultimately distracting, for such a programme can easily become a travelogue, where an attempt is made to explain Greek science in terms of a landscape rather than discussing its intellectual impetus. In this respect science does differ fundamentally from art, which explicitly deals with images and their intuitive enjoyment.

Mr. Burke's *Specials* have been another BBC presentation of science. The format demanded a studio audience participating in a fast, high-pressure explanation of a scientific or technological development and some exposition of its possible consequences. The programmes were universally disliked by science writers and critics. There was indeed gross oversimplification and a deliberate injection of sensationalism. Nevertheless, it is difficult to avoid the conclusion that, although what the audience gained from the programme might not have been what scientific purists would recognize as a true understanding of a scientific development, they received an emotional, intuitive grasp of some facts of science and its possibilities, which perhaps is all that could be hoped for.

Further along the same road is the historical attempt to place a scientist in the perspective of his own times.[51] This is a hazardous undertaking, since there appears to be little connection between a man's scientific excellence and political or social wisdom—possibly the reverse. In terms of information transfer, the viewer has to cope with a two-stranded argument with the predictable consequence that he will despair of the science and concentrate on the emotional side.

Back to earth, one must praise programmes, such as *Tomorrow's World*, that successfully limit their objectives to a narrow, usually technological, view and show something of the actual experimental, operational side of applied science. It can be argued that such presentations contribute little to the imaginative, romantic side of science and do not illuminate the underlying considerations of scientific research. Yet they do emphasize a point which tends often to be overlooked: that at some stage the embodiments of science are operational, functional things, and that for the vast majority of people these constructs will be the only visible achievements of science.[52]

What television cannot do effectively, despite its great technical advantages, is to convey ideas and concepts. Perhaps there is some fundamental difficulty in that no amount of visual explanation and dressing-up in any fashionable representational form will enable an idea to be understood and appreciated without the mental struggle of acceptance, which in turn takes time and effort. It also needs the ability to stop and go back for a second look at aspects misunderstood or implications missed. No public television broadcasting system can cope with this need, and will not be able to do so until the necessary hardware for video recording equipment becomes cheap enough for general domestic use.

There is bound to be scepticism about any means of communication which is ultimately governed by a perceived need to trawl the largest numbers and which in lean years will therefore severly curtail any experiment which departs from proved successes. Nevertheless, the television presentation of science does appear to have far more potential advantages than can be expected from the written media. Newspapers and magazines have become set in their ways of presentation and ideas. Increasing financial pressures have concentrated the efforts of management and editors on the need for technical developments which would make survival possible and have left them little energy for a thorough examination of possible developments in topics and style. Because television is a relatively new medium, its tradition are more fluid, and, at least where commercial pressures are not unacceptably heavy, there are still far greater opportunities for experiment, both in topics and treatments.

The main strength of television is its ability to use the most up-to-date electronic techniques to reach the deepest—essentially non-verbal—levels of communication. This in itself is a temptation and provides an opportunity for misuse, for the impact of television is often not open to intellectual verification, and in the hands of fallible humans it will inevitably be misused. But there are degrees of inaccuracy, and it is arguable that a small amount is of no great consequence provided that the final result justifies the method. The visual aspects of the medium have certainly overcome one of the chief problems of the printed word by forcing scientists to account for their actions in public in such a manner that the scientifically untaught or uncaring should benefit, or at least be interested.

Provided that the canons of good presentation and interesting programme do not calcify into immovable categories, there will inevitably be great changes in choices. It is at present commonly and unquestioningly accepted that science must be humanized, or at least presented in terms of debate and geographical juxtapositions. In fact, there seems no reason why television should not return to an up-dated version of the Victorian illustrated lecture, once producers are persuaded that such a programme would not result in a mass flight to the other channel. Nor is it self-evident that an audience which demonstrably enjoys the intricacies of war strategy or the complexities of economics would be completely out of its depth if confronted with the basic operations of science devoid of any social or economic relevance. Successful examples already exist in the broadcast of lectures from the Royal Institution. Names such as Bragg, Attenborough, Porter, and Laithwaite have become synonymous with enjoyable presentation of both the life and the physical sciences. The need for a lecturer of appropriate quality and personality can therefore be met. But, despite the honest if misguided lobbying of vested interests, changes will occur only when the general ideas of the communicators in television turn towards a new way to explore, prompted by pressures of public opinion, true or imagined, or merely because the old ways have become boring and new ones need to be found.

One must also hope that in future some solution will be found to television's greatest difficulty in this area: the inability to discuss the ideas and logic of science. It might be argued that these present insurmountable difficulties, since they require lengthy and patient study and the slow maturing of responses to new notions and processes of thought. This argument is old-fashioned and does not take into account the newer results of educational psychology, which have clearly demonstrated that ideas, often of great complexity, can be transmitted to those who would not normally be regarded as able to accept them. It can be argued that if the transmission of knowledge changes the recipient operationally, the learning process has taken place, and therefore that the essentials of the idea have been conveyed. Because there is little communication between educational psychologists and television producers, and because the ideas of the former would in any case have to be greatly modified before they could be applied to broadcasting, there will inevitably be considerable delay before any results can be expected in actual programmes. There seems nevertheless no reason why eventually the new educational ideas should not become as much a part of the visual communications of science as are today the views of the pundits in the studio.

A development which would allow television to present ideas successfully, would also allow the full flowering of what would be the most important contribution that television could make to the communication of science: the ability to demonstrate the social consequences of scientific endeavour and the true relevance of science to political and economic decisions. For all except scientists it is not a particular piece of research that matters, but rather the final results, based upon social, political, and moral considerations in combination with research results: in other words, the price society is prepared to pay for particular true or imagined benefits. The juxtapositions that are necessary to illuminate this domain can be achieved in the printed word, although they may require a large and difficult leap of imagination from the reader. In radio, the need still exists. But television can show relationships in a tangible way that is understandable to all, and these relationships must ultimately depend not only on particular research results but on the larger patterns of scientific development and the ideas of its practitioners.

Because of its visual impact, television could become the chief channel of communication, not only from scientists to others, but also from the centre of society itself to the scientists. This would be a remarkable achievement. The traditional pressures on science arise from government or industry, rather than from beliefs about what the community requires. Television could offer a means of establishing the needs of the community at large and of conveying them to the scientist. This would give television a direct voice in the conduct of scientific affairs. One can but hope that those who are directly concerned will have sufficient reason, humanity, and humility to do no worse than their predecessors.

6

Big science and concerned science

In a Victorian book on chemistry, Professor Pepper, lately professor of chemistry and honorary director of the Royal Polytechnic Institution, author of *Various works for youth*, etc., lovingly described a lecture demonstration of oleography, the phenomenon oil forms patterns on water.

> ... a box with a glass bottom was filled with water, and the limelight placed underneath the box. On throwing a spot of liquid, giving a cohesion figure on the water, the figure, more or less definite, was exhibited on a tracing paper screen placed above the box. Even with a candle underneath the box, the figures were visible ...[53]

Incidentally, this experiment could not be repeated with most household oils used today; they contain silicone and the figure will not form.

What would Professor Pepper, were he still alive, make of this?

> The third reactor ... [is] called the Fast Flux Test Facility; it will operate at a very high neutron flux ... [it] will cost about $100m and will operate at a power level of 400 megawatts ... the Commission has taken the first steps toward construction of one or more liquid-metal fast breeder demonstration plants ... the full-scale liquid-metal fast breeder reactor of the 1980s will be rated at 1000 megawatts.[54]

He might well be frightened by the scientific and technical ideas such a description implies, but he would no doubt consider that a little background reading would soon allow him to catch up. Undoubtedly it would. But the $100 million cost would be more than Professor Pepper could imagine, especially in terms of science. It belongs to a hierarchical level outside his experience, and what is true for Professor Pepper is arguably even truer for the scientifically unendowed of today.

This difficulty suggests two more problems in communicating science: size and familiarity. Most of the scientific items that appear in the media concern 'big science', that is, enterprises where the required funding, the product achieved, or both, are on a very large scale. Apart from medical advances, 'little science', where both the achievement and the tools required are on a small scale, only infrequently merits notice. Gigantism introduces the problem that the system, whatever it may be, assumes a life of its own in which scientific considerations play only a part, and the size of the enterprise makes it difficult to compare it with anything within the average reader's or viewer's compass. The vital personal nexus is lost.

Large national and international enterprises are often the subject of debate, but few of the discussions include a deep examination of scientific or technological factors. NASA, *Concorde*, or the newest atomic power station will be debated in a climate of national or ideological emotionalism which is concerned with prestige or the unacceptable face of capitalism, the environment, resource utilization, or economics, paraphernalia whereby democracies assure themselves that despite everything the process of participatory decision-making is in full swing. Within the domain of thundering newspaper leaders, science is scarcely ever mentioned. Many will assume that the argument is hardly about science or its utilization, but is rather a jousting tournament between contemporary pressure groups. Yet even if science were brought into the arena its contribution under present conditions would be only insignificant. In order to make effective statements, it would be necessary to find a public which could not only understand the scientific problems underlying a particular debate, but could also make reasoned choices between conflicting scientific evidence. In such conditions the idea of the concerned scientist is stripped of most meaning since it assumes an ideal state where a knowledgeable public listens to scientific opinion pointing in a single direction; indeed, in practice both scientists and public are divided in their views.

The effect of the very large science-based enterprise based on scientific advances is to repel scientific explanations. It does this by reason of its inhuman size, which places it outside the imaginative power of most people and transfers it firmly to the domain of Them, be it governments, authorities, or organizations, which, it appears, are governed by laws and reasons outside the common compulsions. Whereas a Victorian interested in science could set up his oleographic equipment or other experiment with home-made apparatus, to operate a cyclotron or a small oil refinery is completely beyond the scope of any hobbyist. Yet again, therefore, one has to return to the argument that the existence of big science and its technological consequences have effectively removed science not so much from public understanding but rather from

Big science and concerned science

the necessary nexus of involvement which an interested observer would be expected to evince in the workings of science.

On the other hand, argue the protagonists of easy science for everyone, why not point out how much science and technology has done for everyday life? Why not rejoice in the existence of plastics which wipe clean, washing machines which save drudgery, efficient and comfortable motor cars? Once the public's attention is drawn to all these components of the gracious life, surely they would be only too willing to make the mental effort required for their understanding? Noticeably, this argument is spelled out most often by eminent scientists at times when the grant situation is becoming tiresome, or by equally eminent industrialists considering the wisdom of increasing prices by ten per cent. The careful listener will readily note an overtone of emotional rejection, a plaint that all efforts and travail have been so little understood and rewarded.

This argument does unfortunately show an almost total misunderstanding of public attitudes. For one can hardly complain if, after the expenditure of vast sums of money on advertising and other emotionally guided information, the public quite rightly assumes that the miracles of science are indeed not only here to stay, but that they belong to them as of right, arguing correctly that in most cases they were brought about unasked and, ultimately, for the pecuniary benefit of some individual or group. The public's emotion is not therefore raised by the existence of washing machines, but only by the fact that they break down and are difficult to have repaired; not by the existence of motor cars but by the rising price of petrol. Having paid its money, the public requires the miracles of science to work without further effort and will not tolerate a didactic approach to objects for which it has already paid the market price.

A further consideration of the size factor suggests that it also relates to the number of people involved. A scientific or technological development affecting a small number of people has quite a different social value from one affecting a large number. This is a trite enough observation, but one which was certainly not considered by early writers about the sciences. A Jules Verne or an H. G. Wells could, together with their public, dream about motorized transport or flying machines, and look forward to the day when the mental and aesthetic horizons of all would be expanded by the ability to travel quickly and safely over long distances at will. The dream lasted while only a small proportion of the population could afford motor cars or air travel. But once the average family had 1·35 cars, the other side of the argument became too evident in the miseries of accidents, traffic jams, fume-laden streets, crumbling old buildings, and energy crises translated into higher petrol prices. Similarly, although the early exuberance of the chemical industry was soon countered by the passing of the Alkali Acts, there can be few observers who would fail to notice what large-scale industry has done to its surrounding countryside. Thus too many miracles affecting too many people viewed close are no longer reassuring. On the contrary: present danger foreshadows future disaster because the system has grown too large, because the general assumptions of the past are breaking down and the miracles are crowding in too fast for comfort. Nor is it entirely convincing to argue at this stage that only science and technology are capable of ameliorating the present mess, for the obvious reply will be: Why has it not done so in the past? One is inevitably driven to the conclusion that for society science is only one of many factors in a given situation; we encounter what has been called the multi-dimensional problem.[55] Another way to the same conclusion is the assertion that science is not neutral,[56] which has indeed become, in one or another version, the generally accepted point of view of those examining problems at the meeting-points of science and society.

If, however, we accept that the interrelationships between science and society are complex, we must also take into account the population density of that society. One man's death next door is more tragic than the immolation of thousands on the other side of the globe, and in the same way the presence of scientific or technological artefacts in one's own back garden has a greater impact than the same objects in the middle of some scarcely inhabited countryside. Science and its products have become physically too near for our emotional and sometimes physical comfort, and what previously could be enjoyed at a distance as an idea, has become a tangible, threatening presence.

Those with long enough memories can contrast the euphoria of science in the post-war years with the present disenchantment. Both were based on false premises, but in each case the effect was strong and real. During the Second World War, what was thought to be science, but what was effectively engineering, contributed greatly and obviously to the victory of the allies. After the war, the public expected that the wisdom of the scientists, together with the applications of their inventions, would usher in a new world of plenty, just beyond the horizon. Had not DDT put paid to malaria, penicillin to tuberculosis and venereal diseases, and did not the power of the atom promise energy supplies at virtually negligible costs? It is often forgotten that most of these promises have in fact been fulfilled. DDT and the green revolution have multiplied the world's food supply, and medical science has eradicated vast swathes of disease. All the science-based industries have made great strides, and, compared to their nineteenth-century forerunners have minimized their interference with the environment. What spoilt the idyllic picture was the realization that the promises of atomic energy could not be fulfilled upon the terms promised, and—perhaps more important—that the very successes of science and medicine resulted in an increase in population. Suddenly and unexpectedly, large number of people began to clamour for the same *douceur de vie* which inhabitants of the developed countries already regarded as their birthright.

Rising standards of living, rapidly increasing populations in the less fortunate areas of the world, the realization that material fortunes must be paid for in terms of social cost, and the general feeling that the system had become too complex thus brought about the growing disenchantment with science and all its works that has become the modish expression of the late 1960s and the 1970s.

It is arguable that this phenomenon is healthy, since the public quarrel is not with absolutes but with relative value-judgements. The disenchanted are almost entirely the products of western middle-class affluent backgrounds, who are not so much worried that their own living conditions are too agreeable as that the existence of less favoured populations does not allow them to enjoy their own comforts with a good conscience. Yet one cannot disregard the quite justified political element in the current anti-science feeling. The increase in size in every systematic human activity, be it science, manufacturing, or trade, has led to the emergence of new constellations of power which find only inadaqute expression through historically sanctioned outlets. The result, as would be expected, is that superficially observed politics and debates are frequently out of step with reality. Generations which thought to bring about the new Utopia through the manipulation of the existing system find that, although victory is nearly won, it is a Pyrrhic one, for the substance has escaped. It is a matter of record that most new, aggressive industrial and commercial power bases owe their existence and development to aspects of scientific discoveries: the great multinational companies are not in woodcarving or rug weaving, but rather in oil, chemicals, electronics, and communications. From this it is argued that science and technology, by helping the growth of large oligarchical systems, has reduced men to the status of an insensate cog in a machine, and that science has helped in the dehumanizing process. On any objective view the essence of this argument must be accorded reality, despite managerial blandishments, for no wages, day nurseries, or music while you work can compensate for the utter tedium of most modern manufacturing processes.

But these are long-range arguments which fit ill within an instant communication system. The feelings concerning shifts of political power, the vulnerability of the individual, the social cost of material progress, and the enormity of the system as a whole come together, and the blame is placed on the hapless scientist who contributed largely to the emergence of the world as it is today—as indeed he used fondly to proclaim. The scientist emerges as the new bogeyman, the evil shaman in thrall to capitalism or multi-national enterprises or whatever, despoiling the countryside for the creation of useless and dangerous wealth for a favoured élite. We should note that these arguments are propounded with any degree of sincerity only in the western, developed countries; it is only there that communes flourish, having as their aim a return to bio-organic farming, vegetarianism, and the rejection of a materialist society. An intelligent inhabitant of a communist state will still speak with justifiable pride about the rapid industrialization of his country and the rise in living standards that greater efficiency brings. Similarly, in developing countries, where the national standard of living at present approximates to that of the communes, aspirations reach towards the ownership of a motor car, or at least a bicycle.

Although the present-day critics of science tend to proceed from a neo-marxist or romantic point of view,[57] one should remember that they enjoy respectable historical antecedents. The nineteenth-century poets in both English- and French-speaking countries had little love for science. Lest one should equate dislike of scientific and technical progress with dreamy romanticism, it should be recalled that the first Clean Air Act was passed by Parliament in the thirteenth century, forbidding the burning of sea coal while the House was in session.

Critical attitudes to science and technology found a ready echo in the way science is written about. On one level, the science journalist, if he is good at his craft, must mirror the concerns of the public, and in a sense must be a catalyst for it. As the 'God-how-marvellous' era of science communications passed, a change in direction was in any case due, and it would have been less than human if the general area of the social consequences of science, usually expressed in extreme terms, had not been welcome. There is another strand of the argument. Science journalists can become bored with their job, and a succession of scientific miracles, when they have to be presented day in and day out, tends to strain human patience. Social consequences are heady stuff, for by their very nature they deal with humanity—or indeed Humanity: a much more readily understandable concept.

From our point of view a more significant aspect of the critiques of science is the almost total confusion between science and technology. One cannot but argue that this reflects rather accurately a similar confusion present in the public mind. In a sense this is no more than going back to the days before the Industrial Revolution. It is due, one can suggest, to a combination of generations of bad teaching and misleading public communications. The problem is genuinely complex, but those studies that have been carried out tend to show that the links between science and technology are tenuous and exist most strongly in the trained minds of those whom science had educated to carry out technological developments.[58]

The laxity of the fathers is visited on their children. In the public mind, reporting science is tantamount to defending technology, and the intellectual beauties of the phase rule rest uneasily under the sulphurous fumes of mineral extraction. On the credit side the polemicists have achieved a wide, general acknowledgement of the idea that science and technology are not activities divorced from all else, but are firmly embedded in the structure of society, even if that structure contains the less pleasant aspects of power politics and ideological imperatives.

Classically, science may be regarded as a morally and politically neutral activity. The scientist's interests are limited to the results of his experiments, and his philosophy is a direct and rational consequence of them. If such a position were ever tenable, it is certainly not so under contemporary conditions. Jean-Jacques Salomon[59] suggested the term *technonature* for the area where the interests and attitudes of scientists are inextricably mixed with those in authority. The term was coined to parallel Galbraith's *technostructure*,[60] in order to define those who by dint of their technical knowledge participate in the decision-making of industrial systems.

Nor have some scientists been laggards in appropriating the critique of science argument. Their protestations go well beyond such matters as the elimination of mercury from fish for public sale; nothing if not thorough, they take arguments to their ultimate logical conclusions and prophesy, with the aid of computers, the imminent end of the world.[61] A good apocalyptic story always has a good Press, and it is inevitable that the public, to whom the computer represents the nearest thing to a *deus ex machina*, should be duly impressed.

Big science and concerned science

Science stories for public consumption inevitably follow, at least in general terms, what the public is believed to want. In addition there is the all too understandable journalistic precept that news of actual or potential disasters will be read; optimism might be good for the soul, but it does not sell newspapers. Thus there is a natural tendency to emphasize the limitations of science in action, or—in plain language— the lure of a good scare story is seldom resisted. There is also a common-sense reason. If a reader can be persuaded to read about science through the agency of a scare story, he might acquire the habit and might gradually become interested in wider aspects of science. Technically, the writing or visual presentation of a socially oriented science is far easier than the discussion of concepts and is of far more immediate concern to the reader. To warn against the possibility of an explosion in a nuclear power station is socially laudable even when no such explosion had taken place and even if the possibility of such an event is remote, for should it happen the consequences would indeed be tragic. But a warning does not require an understanding of complex and still-debated statistics, or of atomic theory and high-pressure engineering. To warn against heavy-metal poisoning is valid even if the mechanism of such poisoning is in doubt, the statistics equivocal, and the scientists themselves without a common voice.

Despite the bias of the activities of socially relevant science communicators, one should not forget that their arguments have done a great deal of good. They have drawn attention to facts of contemporary science, technology, and industry which should not remain secret; they have provided the opportunity for some excellent investigative journalism which is at the very core of the craft; and they have punctured the complacency of both scientists and their employers. It is no longer sufficient to shrug off the possible consequences of a scientific discovery or a technological construct, hoping that any ill-effects will quietly fade into the background. Neither money nor governmental power can win for its protagonists immunity against the searching, if cynical, eye of the science journalist.

In the long term it is most unlikely that the current views and techniques of presenting science will be any more lasting than previous ones. There is little doubt that as time goes on more and more fundamental questions will be asked and answers to them attempted. Most of the answers will be wrong, but among them will be some that in the event will catalyse some new idea that will be regarded as significant. Even in the sciences proper, contemporary communications go well beyond the rigid limitations of the academic departmental structures of the past, so that new interdisciplinary areas of research and discovery are increasing in importance. In the same way, the reporting of science will also tend more and more to view, not only science itself, but also its technological, economic, political and environmental consequences. Some of the stridency of current opinion will subside, but other views will take their place, for in humanizing science its effects on the public good are of the first importance, and genuine concern with the public will be required rather than intellectual pretensions. Such processes have already occurred in other branches of scholarship; the historian of today is greatly different from his namesake of even fifty years ago; a botanist has gone beyond the stage of collecting data; and the geographer can legitimately view himself as part either of sociology or of the physical sciences. Science journalism has gone through the stage of fact broadcasting and subsequently through the stage of discovering that facts alone do not contain all the answers. It requires little power of prophecy to suggest that it will next partake in the process of integration, of considering the web of consequences emanating from a scientific fact. In this way it may truly justify its claim to be an important branch of scientific communications.

7
Two-way communications

Communications between scientists and others are usually considered to be a one-way process. Traditionally, scientific information is thought to exert influence in three directions. First, underpinning technology, science is considered to be the main initiator of material advance in the exploitation of new materials and processes. It has been shown that a simple one-to-one relationship is probably untrue, although the exact nature of the information transfer is not yet understood.[58] Secondly, science is also one of a number of factors determining decisions made by government, industry, and other centres of power. This role is the subject of a voluminous literature, some of it forming part of the debate about national science policies.[62] Centres of power have their own translation and screening facilities which allow them to view science, however imperfectly, in relation to their particular preoccupations. Thirdly, there are communications about science to the general public, which for the purposes of this argument is assumed to be without prior scientific knowledge. Such communications may be by the printed word—in both editorial and advertisement matter—and through broadcasting.

In government and industry, there exist mechanisms for providing the scientist with guidelines—or indeed orders— for his future activities through the utilization of political power. Since both government and industry are explicity charged to use science for their specific purposes, the guidance the scientist receives is what the Americans call mission oriented, that is, it has to fit within a given perception of the future needs of his employer.

There are no corresponding mechanisms between the general public and scientists. Arguments for providing more scientific information range from the moral and cultural to the political and economic. In the opposite direction one finds a number of general injunctions for greater social concern, but apart from a few individuals or groups (more often than not regarded as cranky by the average scientist), there are few indications of how the scientist may act for the public betterment in a professional capacity.

In the same way that a translation is needed to communicate the ideas of science to non-scientists, a new language will become necessary to convey the needs of society to the scientists in ways they can readily interpret. Only in this way can the scientists, *qua* scientist, become a professional concerned with the problems of society at large, and make a positive contribution.

It could be argued that the scientist, who outside his own field of specialization is on par with other members of the general public, needs no such special treatment. There are already those who voice the concerns of society, and the scientist has only to listen carefully. Such an argument does not face the realities of the situation. Elected or self-appointed consciences of society hardly declaim with a common voice. Who is one to believe? Which political, ideological, or group pressure should the scientist take into account? But perhaps more important is the actual rather than the ideal response of most scientists to these blandishments. If scientists have evolved a jargon of their own, so have sociologists, politicians, and special pleaders, and the effect on the scientist is predictable; he, too, is affronted by phraseology he cannot understand, mistrustful of ideas only imperfectly glimpsed, and in the outcome he will turn away from the conflict of emotional debate to the well-understood logic of his own discipline. As a semi-quantitative proof of this assertion, one may observe that, although their noise output is considerable, the membership of all the societies and groups catering for the socially concerned scientist is minute.

But, suggests a further argument, how can society define its needs when so much of what science has discovered is secret? Indeed, how can the socially concerned scientist work on problems of relevance if he is unaware of the actual situation brought about by the clandestine use of science and technology? This argument often confuses true secrecy with information scatter. In relatively open societies large areas of science are not in the public domain, but it is surprising how much is hidden, not because of authoritarian command, but simply because the information is not where the enquirer assumes it to be.[63] An obvious example relates to the presence of heavy metal pollutants in the atmosphere and in food.[64] For a long time physical scientists were unaware of the dangers posed by these materials, and government departments took even longer to be persuaded of their reality. The reason was not so much that data relating to metal poisoning was secret or suppressed, but simply that it appeared in journals circulating to physicians rather than to scientists. Governmental and industrial secrecy undoubtedly exists, and it would be difficult to imagine a state of affairs where it did not, irrespective of the political hue of the authorities. But a great deal of information is hidden rather than secret and could be found, given enough time, persistence, and intelligence.

For the science journalist an important role is to contribute to the development of a new language, not only for informing the public about science, but in making the scientist aware of the public's needs. In this role he will have to employ the best traditions of investigative journalism: not through the crudities of the cheque book, the hidden microphone, or the corrupt official, but simply by using his intelligence and his knowledge of how scientific information is processed to collate and synthesize the available material into usable data for both scientists and the general public. Such a view of the future emphasizes one aspect of science journalism which has been little discussed by journalists or scientists: that in a number of important ways the scientist and the journalist conduct their affairs similarly.

One of the most important functions of both is to validate a result, that is, a piece of information. The scientist repeats his experiment: the journalist seeks collaboration from different sources. Both attempt to spot weaknesses in argument and both pay attention to where the correct argument leads. Both develop their hypotheses by collating available information and both investigate the validity of a hypothesis under different experimental conditions.

Future developments

It would be seriously misleading to assume the existence of a homogeneous general public; needs and interests are too different. One must also exclude from the general public those who, although not scientists, receive scientific briefs as part of their work. It is unlikely that the managing director of a large company or a politician in charge of a department of State lacks scientific information. It is possible that such information is contradictory, unreliable, or not acted upon, but that is another question, unrelated to problems of general communications. In this domain the function of the science journalist is investigative; by pointing out sins of omission and commission, he will assist in developing the existing channels of communication to take cognisance of problems which ultimately have to be solved by political or economic action.

Other groups of audiences or readers do not enjoy built-in channels of scientific communication. They differ according to innate abilities of dealing with information and educational accomplishments. The responses of a reader of, for example, the *Daily Mirror*, may well be different from those of a *Times* reader. Yet in all groups there seems to be an unfilled interest in at least some aspects of science, as envinced by the very large television audiences for programmes which would superficially appear to be highly specialized.[65] Although one can argue about both the quantity and quality of scientific information conveyed to these groups, it seems probable that at least on occasions the problem is not too little but too much scientific advice, which ends by confusing the individual.[66] The phrase 'information overload' has been coined for this situation, which is exemplified by the contradictory advice given to expectant women by doctors, clinics, and various publications.

Ultimately the solution of how to convey the essence of scientific thinking to non-scientists must be a matter for education. Whether the reforms of curriculum and methodology in schools will be the answer, only the future can tell; but it would be safe to suggest that whatever the outcome it will present a long-term phenomenon. The science journalist must deal with the present, when the onus is unequivocally on his shoulders.

Irrespective of the particular technique of communications required by the different recipient populations, the science journalist must consciously decide what sort of science will be of greatest interest to his readers or viewers. One can indicate three approaches, according to which aspect of science it emphasizes: discovery, social, or humanistic.

Discovery-oriented science journalism is primarily concerned with the facts and ideas of one or a series of scientific developments. The coming of age of tectonics is a good example. Discovery in science is still a potent source of interest and awe, providing an increased ability to learn more about man and his surroundings. It assumes on the part of the audience an ability to enjoy ideas, to rise above the utilitarian, and to be interested in what after all is a large part of present-day cultural activities. This approach to science journalism is currently out of fashion, except perhaps on television where the visual qualities of science can be harnessed to the presentation. Nevertheless, in the same way that the discovery aspect of science was much enjoyed in the past, it seems likely that as fashions change it will once more be greatly appreciated. Perhaps this aspect of science for non-scientists is most dependent on educational attainment, not in its fact-hoarding aspects, but rather in attaining an ability to appreciate the world of ideas and the elegance of patterns.

Social concern is the currently most widespread approach to science journalism. At its centre is a belief that the social consequences of science and technology are more important than science itself. Although it can spill over to anti-science, as its best it does indicate a concern for those who become the unwitting participants in social processes originating with scientific discoveries. At its lowest level one can regard it as the soft option, for it is obviously easier to describe people's feelings or a changed countryside than to explain the often complex factors involved in the embodiment of a new technology. Unfortunately this approach coincides rather well with educational developments, since there are serious indications that students' preferences at university level reflect not so much a lack of interest in science but a lack of interest in subjects where hard work is expected.[67]

As a corrective to the previous all-too-uncritical acceptance of scientific advances, the new approach along societal lines has been valuable. Following the swing of the pendulum, the reaction has arguably been too extreme. Thus it seems reasonable to guess that the pendulum will swing again and, although the consequences of science will still be regarded as an important aspect of scientific information, they will be regarded as much less central.

Perhaps a synthesis will arise along what one might define as a humanistic approach. At present this is usually manifested in describing science in historical terms, that is, in terms of the scientists themselves and their contemporary environments. If it could be sufficiently developed, this approach would offer a satisfactory amalgam of all the significant factors comprising a scientific discovery and its aftermath.

Any scientific development depends on the people making it, and the scientists themselves must be influenced by both the scientific and the cultural environments they inhabit. In addition, they must become aware of the social consequences of their work, however many steps removed. Thus a system, or a pattern of interactions, showing the place of the individual and his work within a social framework must ultimately be the correct and most accurate treatment of science. Such an approach would also meet the requirement of showing science, not as a self-sustaining activity in a social void, but as a cultural paradigm deeply embedded in the ideas, needs, and, on occasion, myths, of contemporary society.

The segregation of scientifically and humanistically inclined students is a phenomenon that owes more to administrative and educational convenience than to deep philosophical reasoning. The arts and the sciences have more in common both in their patterns of work and of thought than is commonly realized;[68] they may deal with different aspects of reality, but their common preoccupation is the enlargement of man's consciousness of both himself and the natural world. Once this is taken for granted rather than forming a subject of contention, the science journalist can perform the valuable function of a true bridge-builder by illuminating areas of joint concern.

Presentation

Improvements in communicating science must be contingent on improving mechanisms of presentation. Looking at the traditions of newspaper production, there is little reason to be optimistic. The economics of newspaper publishing are complex, but it needs little expertise to notice that the greatest economic pressure falls on the popular end, and in the battle for circulations there seems little hope that science, essentially a fringe subject, will receive better treatment. With slightly more economic and—dare we say it—intellectual elbow-room, it would be relatively easy for at least some newspapers to generate a new look at science. Only slight shifts in attitude are needed: for example, an acknowledgement that the purveying of news by newspapers is gradually being eroded by radio and television, and that seeking complete up-to-dateness is chasing an *ignis fatuus*. A considerable portion of every newspaper is no longer concerned with news but rather with what is referred to as background information: an explanation, and possibly an illumination, of the causes underlying the day's events. There seems little reason to doubt that a gradual realization of this phenomenon will dawn on newspaper editors and proprietors, allowing a far greater range of subject-matter and the occasional treatment in depth of significant developments—in science as in anything else.

Yet another factor is the real or imagined need to preserve a stylistic straitjacket, to conform at all costs to a set of empirical rules. Admittedly, in order to maintain any degree of cohesiveness any publication must have a set of house rules in matters such as design, spelling, and punctuation. There is also a natural tendency to evolve a particular style and internal logic applied to all printed matter on the page. Yet wise editors are aware that internal rules are guidelines rather than immutable barriers and that too much reliance on them eventually leads to the detriment of the publication. It would thus be possible to expect newspapers to show a certain flexibility to accommodate the divergent needs of the science specialist.

Nevertheless, none of these minor developments in favour of science reporting can have much effect on the inbuilt mechanism of newspaper production. All papers are subject to the immutable demands of a tight time schedule, of the necessity to fit together the complete issue in a limited number of hours under conditions of stress, choosing a tenth of the available material on grounds that are essentially empirical. In the disciplined chaos of a deadline it is inevitable that personal preferences, idiosyncrasies, and power politics assume an importance they would otherwise lack since there is physically no time to bring in any countervailing mechanism. It is this necessity, rather than some dark conspiracy, that militates against most papers having any sort of forward planning which would allow them to consider topics comprehensively and at length.

Despite these arguments, newspapers themselves have demonstrated that they are perfectly capable of imaginative forward planning if it is in their interest: witness the special features, heavily dependent on advertising revenue. It is only a small step from this idea to a planned, ongoing presentation of material, tied to the daily news in only the most general way. Such a change of attitudes will have to come from those who work on newspapers, since it is only in the context of their daily preoccupations that it can find a satisfactory embodiment.

Should a planned presentation of material become the accepted practice, it could be comparable with the current treatment of other cultural activities, for example, films or the theatre, which are already only 'news' because they have been so defined. A changed attitude would emphasize once again that there is nothing in science to exclude it from the general cultural activities which are already accepted as part of a newspaper's sphere of interest.

It is only realistic to accept that all changes in presentation and emphasis may have economic consequences. Most national newspapers are already walking an economic tightrope. Among magazines which have a right to be called popular, the same economic considerations are present. Nevertheless there are special factors in their favour: a relatively greater cushioning of advertisements and, generally, a fairly stable list of subscribers who, fortunately for the publishers, pay their subscriptions at the beginning of the year. Unfortunately the editorial and mechanical production of a magazine represents a much smaller system than a newspaper, and the natural tendency is that once a profitable formula has been found, it is not to be abandoned except for very good reasons. It has been clear for some time that the British market, for example, is unable to support more than one general science magazine: a greater interest in science must therefore emerge from the widening horizons of the specialist journals. This in turn is contingent on what the publishers of specialist magazines perceive to be of interest to their particular publics; that is, on how far the specialist in one area is likely to broaden his awareness. It is certainly the case that the spectrum of interest among magazines serving scientists (as opposed to research journals) is now much wider than, say, ten years ago. Whereas a magazine designed for the general interest of chemists or physicists, typically published by a learned society, would have contained little more than institutional news and scientific snippets, today's equivalent will carry a wide range of articles, including ones from industry, politics, and economics. There is also the surprising viability of other journals, still addressed mainly to scientists, covering a topic area (for example, energy) to which a number of disciplines contribute.

A significant tendency among scientists working on some of the most exciting current problems is the formation of teams. These are essentially multi-disciplinary, each subject specialist contributing his own expertise to the solution of a common, many-faceted problem. The result is a synthesis of specializations for the common good. Extrapolating this trend, it would not be unreasonable to hope that the lead given by the scientists may be followed in a more popular domain, although undoubtedly this will take a long time.

Television and radio, at least in the public sector, are less exposed to the pressures that affect printed publications, principally because the cost of most science programmes is less than that of others. Television especially commands audiences of reasonable size; the numbers are at least sufficient to safeguard the development of the programmes. Thus television presents the most hopeful future for improved communications in science, underscored by the visual qualities of science itself. But large areas of science abounding in rich and exciting ideas and concepts are not *prima facie* visual, and some method should be found for their presentation. Television, in parallel with newspapers and their obsession with house style, has its own shibboleths, not only in terms of visual qualities but also concerning the visual effects it considers can be best projected. One could be charitable and suggest that these notions are ultimately derived from television's visual ancestors, the Hollywood epics, and that once the medium achieves maturity a much more objective view will be taken. Television can already present social or political themes through discussion and, irrespective of one's feelings about studio presentation, if the topic is sufficiently interesting the presentation takes a deserved second place. Science can be made as exciting in terms of ideas as politics and therefore could make do with a visual approach in a much lower key. If a Shaw in one era or a Beckett in another can hold theatre audiences spellbound by the eloquence and interest of their ideas, dispensing with virtually all 'business' and action, one might expect television audiences to be able to claim a comparable degree of sophistication and thus accept a similar treatment for matters scientific. A step in this direction has been the televising of the *Controversy* programmes from the Royal Institution, where the audience is encouraged to participate in a debate presenting differing scientific points of view.

All improvements in the presentation of science by the media hinge on the greater involvement of the professional scientist. Time and time again the scientist is criticized for being secretive, unable to communicate, unable to explain, unwilling to share his ideas, and patronizing if not downright rude. Although these difficulties are primarily the concern of the newspapers, they also affect the other media.

There is a great need to establish better channels of communication between the scientific community and the media. In functional terms, this means better relationships between scientists and science correspondents or their equivalents: a relationship based on a common interest, and if not on liking at least on respect and tolerance. The advantages of improved relationships would be equally shared between the journalist and his public, on the one hand, and the professional scientist on the other. In the long run the present ambivalence in public attitudes towards science can be overcome only through a better public understanding of what science can and cannot do; in turn public policies, and therefore the very future of scientific endeavour, are intimately connected with public attitudes. Nor can a professed concern with the social consequences of science be divorced from better public knowledge, especially since scientific opinion is frequently conflicting. Better communications with the public are therefore very much to the advantage of the scientific community, and enlightened self-interest, if nothing else, should make the scientists willing to accept the implicit challenge.

From the point of view of the science journalist, and especially his editor, there should be a clear distinction between instant news—'X

discovers cure for cancer'—and the equally important stream of background information which allows an item to be placed in its correct perspective.

In the past various mechanisms have been tried. Most learned societies and professional bodies issue comments on developments that they consider to affect their members' interests. Unfortunately, such comments are official in the worst sense of that word, and by the time the appropriate council has had time to reflect upon the matter and issue its often unreadable statement, most people have forgotten the original stimulus. Furthermore, learned and professional societies generally deal with one or a very restricted range of disciplines, and it is regarded as poaching should one institute comment on matters which logically fall into the ambit of another.

This leaves the Royal Society and the British Association. On some matters of scientific interest the Royal Society has been known to issue a public comment, but the restrictions operating in all learned societies apply to it as to the others. In addition, the Royal Society, with its many connections with the government of the day and with foreign academies, is very much the official voice of British science, and it sensibly regards itself as capable of more effective influence in private than as a protagonist in public controversies.

Among the original aims of the British Association, the encouragement of the spread of knowledge ranked high. As has already been noted, the Association has begun to publish commentaries on selected scientific and technological developments and the use that society does or can make of them. These publications are based on the deliberations of committees meeting under its aegis. Despite this new endeavour, and despite the limited mixing of scientists and non-scientists at the annual meeting, it is difficult to envisage the British Association acting as a go-between. The Association is held in great affection by both scientists and journalists, but it is only fair to suggest that there is something missing. Perhaps as a consequence of its long and illustrious history, or the limitations of available funds and manpower, the Association projects an aura suggesting that its present preoccupations are probably all it can cope with, and that further tasks would be beyond its undoubtedly sincere and talented officers and membership.

The specifications for a body to encourage exchanges of information between scientists and journalists, to act as a clearing-house of ideas, and to educate both scientists and journalists are reasonably straightforward. It must be completely independent, owing no allegiance to scientific, political, or power-centre ideology. It must not be official, in the sense of having to act as a protagonist for a given point of view. It ought not to be burdened by historical precedent and by implicit ideas of how certain topics ought to be approached. On the contrary: it should have the courage to experiment, to allow its imagination to lead it into new ways, to make its own mistakes. Neither science nor the general public, as represented by the science journalist, form homogenous bodies; controversy and debate must therefore occupy an honoured place in communications between the two.

Meetings should be informal. Both sides should feel able to put forward points of view without fear of appearing ridiculous or of offending actual or potential vested interests. In essence, such a clearing-house should return to the aims and ideals of the local scientific society, although the locus in this case would be intellectual rather than geographical. At its beginnings the British Association described a similar idea in the phrase 'the peripatetic university'. Others have used the expression 'the invisible college'.[69] In each case the desire was to establish an informal but informed meeting-place between people of different professions with common interests.

Arguments against such an idea must include both economic difficulties and the possibility of censorship or manipulation. Clearly, even an informal arrangement must cost money. But if all sides desire it, and are persuaded that their interests would be well served, such moneys could be found. For example, universities and most scientific establishments already employ people to publicize their work and to act as sources of information. The media spend considerable sums to obtain material. It does not appear too outrageous to suggest that, given the will, a system of shared costs could be developed so that individual costs could be minimal.

More important are considerations of censorship or manipulation. The suspicion could easily arise that the scientific community acting as a body might try to manipulate the journalists into particular attitudes to influence governmental and other sources of funding. From the opposite side, scientists might well fear that a cohort of sensation-seeking pressmen would wish to trap them into making damaging statements out of context.

Such arguments can be disposed of by emphasizing that the invisible college would have no monopoly of information and would faithfully mirror heterogeneous scientific opinion. A dissatisfied journalist could easily go outside its ranks to seek information, as indeed he would in any case for confirmation. Nor is there much substance for fears arising from the Press's eternal quest for sensation. A large number of sensational stories arise through sheer misunderstanding rather than through distortion; better understanding and the ability to place a scientific news item in its context are the best weapons against misleading information.

The most important function of such a meeting-place would be to prepare the ground for an item of news by providing the background through a continual exchange of ideas. This would be important, not only for the journalist but also for the scientist; for example, by providing an indication of how others might react to the news of particular developments. The essential value of the invisible college would therefore be in providing a mechanism which would obviate the necessity for the instant reaction: by the time the news broke both scientists and journalists would be in a position to appreciate its importance.

The successful establishment of a new meeting-place would require the co-operation of the best people that each side can muster. In terms of the scientific community, there would need to be known experts in numerous disciplines, balanced by less senior colleagues; in this way accusations of establishment or trendy revolutionary cliques could be countered. Even more important would be the personal characteristics of co-operating scientists. In addition to their expertise they would need to be reasonably articulate, or at least willing to learn the arts of self-expression.

Twenty years ago, the majority of successful scientists would have rejected such a proposition. Today an equally large majority would be only too willing to give of their experience. At the most tangible level, better communications about science would help in creating an atmosphere in which science could again be better appreciated. Although writing about science in the media has no direct benefit to the scientist, it is noticeable that certain topics receive considerably better treatment after they had been introduced into the realm of general ideas; the establishment of the Natural Environmental Research Council is a good example of this.

In order to obtain the largest possible initial response, the college would need the support of a major scientific body, such as the Royal Society. The involvement of such a body would assure the scientific community that the attempt was a sincere move to meet the challenge of scientific communications, while the emphatic independence of the college would be a guarantee that the attempt was a genuinely new departure and not merely a front for conventional methods and ideas.

One should never forget that the communication of science is a two-way process. If the phrase 'the concerned scientist' is to mean anything at all, it must mean a responsiveness to the needs of the society the scientist is living in, both by being influenced in research by societal needs for the future, and by allaying the public fears that are occasioned by conscious or unconscious misinterpretation of facts. The concerned scientist should be able to assess at least some of the fears, anxieties, and hopes of the general public not, as now, through instant responses to particular news, but as part of a continual dialogue with those who have at least a concern for the wellbeing of science.

Whether or not the idea of the invisible college will ever be put into practice, the improvement of scientific communications will be greatly dependent on the improvement in the quality of science writing. The main task of the science journalist, as of the scientist, is to appreciate the significance of a particular fact, to be able to screen the signal from the surrounding noise, and to illuminate true significance in terms that his readers (or viewers) can appreciate. Only in this way will he be able, on behalf of both the scientific and the general communities, to ask the right questions on science, technology, and the needs of society. One is not arguing for a minor sociological treatise every time a science journalist notices a scientific development; but what scientists and the general public can both demand is a rational but concerned attempt to cross the language and thought barriers.

The science journalist need not be a paragon, but he should have an inquiring, nimble mind, a good memory (or a better notebook), the physical endurance of an ox, at least an elementary appreciation of human diplomacy, and the fundamental ability to write in an interesting and grammatical way. His occasional lapses, whether due to haste, misunderstanding, or deliberate misleading by his sources, will be forgiven if he is seen genuinely to be trying to make the best of a complex and responsible job.

We are still a very long way from such a Utopian state of affairs. Much science writing is turgid, verbose, and undisciplined and suffers from sheer laziness. It is so much easier to accept the word of just one source, to look at just one article, or to take over large unedited chunks of a press release sent out by an interested organization. When time or money are in short supply, it needs additional reserves of persistence or desire for quality to make the extra phone call, rewrite another paragraph, or spend another hour in the library. Especially is this so if, on the face of it, better quality does not appear to bring in its wake additional approval.

There exist therefore the makings of a vicious circle, which must be destroyed. The complaint of many science writers that their editors or managements do not understand their special problems is answered by the other side, which contends that their understanding is only too great, and their enthusiasm consequently minute. It is up to the science writers to change this situation. Some will attempt it through bombast, some by the cult of personality (their own), but in the long run the most efficacious method will be the building of good, solid foundations. In the same way that mutual esteem between scientists and science journalists can lead to constructive arrangements, enabling them to accept each other's competence and professionalism in the same cause rather than interfering in each other's work, the proper relationship between the science correspondent and the medium he works for can be established only through mutual respect and confidence.

If the science correspondent is not to be regarded as a shadowy fringe figure, he will have to see to it that his professional expertise and reputation improve. No valid educational methods have as yet been found for producing correspondents; although a number of them have studied science at university level, others drift into the job, learning as they go along. It is difficult to see how a more formalized structure would offer any benefits except by ossifying a hardly ideal situation. There is already an association of science correspondents, the Association of British Science Writers, which in its current activities is hardly more than a luncheon club. It is impossible to see such an organization becoming a professional institute in its own right, aping the older-established institutes by establishing qualifying examinations and so forth.

A more effective mechanism would be that of the market-place. Although the myth of quick hiring and firing still surrounds all the media, in point of fact a surprising amount of dead wood is propped up by false sentimentality. If professional standards are to improve, it will become obvious that lame dogs can be helped over only so many stiles, and after a decent interval the situation has to be faced that some other activity is more suitable for the talents of a particular individual. Such a weeding out of the feebler brethren would have the double advantage of encouraging the others and also of creating the necessary understanding that to be allowed to be a science correspondent is honour indeed.

Two-way communications

There ought to be a far more clearly defined career structure for those who make the grade. The English system has a great talent for placing an individual who is outstanding in one sphere of activity in a prominent position in another; *vide* the successful scientists who create chaos by becoming heads of departments, or the ineffectual personnel officers whose military careers have been a shining example of martial virtues. Why should it be necessary for a successful science writer, if he has greater ambitions, to abandon the craft he is good at to take up another appointment offering better pay and prospects but which only makes both him and his superiors miserable? Why could there not be a career structure for the science writer *per se* in whatever medium he is working which would offer, if not riches, at least a reasonable middle-class standard of living so that a large portion of his time should not have to be used up in devising schemes for earning extra money? This should be considerably easier in larger organizations than in smaller ones, but even in the latter a great deal more could be done. In the past similar arguments were propounded for the professional scientist in industry: that if he wishes to progress, his route must be through administration or some other activity divorced from science. The argument is perfectly valid for those who wish to choose one of the available routes, but why must there necessarily be a choice? Some of the largest industrial companies, whose motivations are no more altruistic than those of the newspapers, have found that it is possible to create a career structure in terms of specialized knowledge which allows both individual and company to derive the full benefit of a talent perfectly honed and effectively canalized.

It has been said that in science everything will be done if people want it enough.[70] Similar arguments can be applied to the communication of science outside the scientific community. Should the general public have a rational view of the glories and dangers of science? Should there be a mechanism to alert the public to both future dangers and benefits? Should intelligent public opinion influence the policy-makers and encourage them to look outside the rank of professional advisers? Should scare stories be neutralized by a general awareness of the limitations of science, coupled with a knowledge that the price of better science is eternal vigilance? Should the scientist be afforded guidelines to the future needs of society in addition to his official orders?

And, fundamentally, should science be regarded as a creative humane activity, an enlargement of human consciousness, no less the glory of an epoch than the frescoes of Michaelangelo or the sonnets of Shakespeare?

And should not the answers to these questions result in science forming part of a social debate before major issues are decided, rather than after?

References

1. James Friday, lecture at the Royal Institution, Summer 1974.
2. See, for example, *Daedalus* **103**, 3 (summer 1974).
3. C. P. Snow, *The two cultures and the scientific revolution* (Rede lecture) Cambridge University Press, (1959).
4. See, for example, D. H. Brooks, *Science, growth and society* OECD Paris (1971).
5. See, for example, A. King, *Science and policy.* Oxford University Press, (1974).
6. *A framework for government research and development*, Cmnd. 4814, HMSO, London (1971).
7. W. De'Ath, *Observer Magazine* (8 September 1974).
8. D. S. Davies, The short term and the long term. in *What kinds of graduates do we need*? (eds F. R. Jevons and H. D. Turner) Oxford University Press, London (1972).
9. T. S. Kuhn, *The structure of scientific revolutions.* Chicago University Press (1962).
10. Sir Herbert Read, *Art Now.* Pitman, London (1960).
11. Salzburg Declaration, European Union of Associations of Science Journalists, Vienna, Österreichischer Klub für Bildungs–und Wissenschaftsjournalisten (1974).
12. See, for example, J. Ziman, *Public knowledge* Cambridge (1968).
13. A variant was quoted by Shelton Bank at the International Congress on the Improvement of Chemical Education (UNESCO), Wroclaw, Poland (September 1973).
14. F. S. Dainton, *Science: salvation or damnation*, 17th Fawley foundation lecture, Southampton University (1971).
15. P. Matthias (ed.), *Science and society.* Cambridge University Press (1972).
16. E. Ashby, *Technology and the academics.* Macmillan, London (1958).
17. Peter Mathias, Who unbound Prometheus. In reference 15.
18. Roy M. Macleod, Resources for science in Victorian England. In reference 15.
19. A. J. Meadows, *Chem. Brit.* **9**, 504, (1973).
20. Morris W. Travers, *The discovery of the rare gases.* Arnold, London (1928).

References

21. W. V. Harcourt, letter to Faraday, 5 September 1831, and subsequent correspondence. In *The selected correspondence of Michael Faraday*, (ed. L. P. Williams). Cambridge University Press (1971).
22. Magnus Pyke, *Chem. Brit.*, **10**, 116 (1974).
23. See, for example, *Charter of the Chemical Society*. The Chemical Society, London.
24. E. S. Coulson, private communication.
25. Malcolm Robinson, *Chem. Brit.*, **10**, 374 (1974).
26. A. Etzioni and C. Nunn, *Daedalus*, **103**, 3 (Summer 1974).
27. S. M. Friedman, in *Science in the newspaper.*, AAAS, Washington (1974).
28. For example, P. R. Ritchie-Calder, *Science makes sense.* Allen and Unwin, London (1955).
29. Norman Metzger, in reference 7.
30. D. Perlman, in reference 7.
31. L. Nathe, in reference 7.
32. D. Fishlock, *Nature, Lond.* **250**, 747 (1974).
33. J. Wood and J. Ronayne, Scientific and technical news periodicals (OSTI Project SI/21/36). University of Manchester M.Sc. thesis (1972).
34. David Martin, O what fools these intellectuals be, *The Times Higher Education Suppl.* (11 October 1974).
35. Report of an enquiry into the use of academic staff time. The Committee of Vice-Chancellors and Principals, London (1972).
36. J. Langrish, M. Gibbons, W. G. Evans, and F. R. Jevons, *Wealth from knowledge*, Macmillan, London (1972).
37. *New Scientist*, **57**, 469 (1973), referring to J. Guttermann *et al.*, *New Engl. J. Med.* **288**, 169 (1973).
37a. A. J. Meadows, *Communication in Science.* Butterworth, London (1974).
37b. P. B. Medawar, *The art of the soluble.* Methuen, London (1967).
38. W. Keble Martin, *Concise British flora in colour.* Sphere, London (1972).
39. Desmond Morris, *The naked ape.* Jonathan Cape, London (1967).
40. Lionel Trilling, *Liberal imagination.* Penguin, Harmondsworth (1970).
41. Jaques Monod, *Chance and necessity.* Collins, London (1972).
42. Victor K. McElheny, in reference 7.
43. For example, Jon Tinker, *New Scientist* **48**, 317 (1970); **63**, 489 (1972).
44. Steven Weinberg, *Daedalus* **103**, 3, 33 (1974).
45. For example, J. D. Bernal, *Science in history*, Watts, London (1954); *The social function of science.* Routledge and Kegan Paul, London (1939).
46. J. S. R. Goodlad, *Science for non-scientists*, Oxford University Press, London (1973).
47. R. W. Reid, *Nature, Lond.* **223**, 455 (1969).
48. A. V. Hill, quoted by Sir John Kendrew, *Chem. Brit.* **10**, 439 (1974).

References

49. The Violent Universe, The Mind of Man, The Restless Earth, The Life Game, The Weather Machine.
50. The Ascent of Man.
51. The microbiologists, produced by Peter Goodchild, BBC TV.
52. R. Baxter, *RSA Jl.* **123**, (5225), 273–86 (1975).
53. J. H. Pepper, *Chemistry, embracing the metals and elements which are not metallic.* Warne & Co., London (1887).
54. Glenn T. Seaborg and Justin L. Bloom, *Scient. Am.* **13**, 223 (1970).
55. Council for Science and Society, Working Party on Information Overload. private communication.
56. H. and S. Rose, *Science and society.* Allen Lane, London (1969).
57. For example, J. R. Ravetz, *Scientific knowledge and its social problems.* Oxford University Press, London (1971).
58. J. Jewkes, D. Stillermann, *The sources of invention.* Macmillan, London (1969).
59. J.-J. Salomon, in *Scientists in search of their conscience* (eds. A. R. Michaelis and H. Harvey) Springer, Berlin (1973).
60. John K. Galbraith, *New industrial state.* Deutsch, London (1972).
61. For example, D. H. Meadows, D. L. Meadows, J. Tanders, and W. W. Behrens, *The limits of growth.* Earth Island, London (1972).
62. For example, J. G. Crowther, *Science in modern society.* Cresset, London (1968).
63. D. Bryce-Smith, personal communication.
64. Peter J. Smith, *Nature, Lond.* **251**, 560 (1974).
65. Nigel Calder, lecture to the Royal Institution Science Writers' Group (Autumn 1974).
66. Council for Science and Society, private communication.
67. D. Duckworth and N. J. Entwistle, *Ed. Res.* **17**, 1 (1974).
68. D. Dickson, *Impact*, **24**, 69 (1974).
69. The term 'invisible college' was first used by Samuel Hartlib and Comenius in 1641, quoted in J. E. Sandys, History of Classical Scholarship (1908), referred to in *Chambers Encyclopaedia.*
70. Sir Peter Medawar, lecture to the Royal Institution Science Writers' Group (Autumn 1974).

Index

advice, scientific, 1–2, 74, 75
Ashby, Baron (E. Ashby), 17–18, 89
Association of British Science Writers, 85

Bank, S., 89
Behrens, W. W., 91
Baxter, R., 91
Bernal, J. D., 90
Bloom, J. L., 91
biology, *see* life sciences
British Association for the Advancement of Science, 20–1, 81
British Broadcasting Corporation, 2; see also *Horizon, Mr. Burke's Specials, Tomorrow's World*
Bronowski, J., 58, 59
Brooks, D. H., 89
Bryce-Smith, D., 91
Burke, Mr., 59

Calder, N., 58, 59, 91
censorship, 82
chemistry and chemical industry, 16, 18
Coleridge, S. T., 19
commercialism in news media, 30
Controversy (series), 80
Coulson, E. S., 90
Crowther, J. G., 91
cultural aspects of science, *see* science, cultural aspects

Daily Mirror, 45, 52, 75
Dainton, Sir Frederick (S.), 16
Davies, D. S., 89
DDT, 66–7
decision-making, 3–4, 5, 72

De'Ath, W., 89
Dickson, D., 91
Discovery, 39
Duckworth, D., 91

economics, 50
education, 14, 16–17, 24–5, 75
energy crisis, 30
Entwistle, N. J., 91
environmental sciences, 50–1, 66; *see also* pollution
Etzioni, A., 25, 90
European Union of Associations of Science Journalists, 89
Evans, W. G., 90

Fishlock, D., 90
Friday, J., 89
Friedman, S. M., 25, 90

Galbraith, J. K., 91
genetic engineering, 4
Goodchild, P., 91
Goodlad, J. S. R., 90
Gibbons, M., 90

Harcourt, W. V., 20, 90
Harvey, H., 91
Hill, A. V., 57, 90
Horizon, 52–5

Industrial Revolution, 17
information, scientific, 1–2, 5, 9–13, 74–5
institutions, scientific, 20–4
'invisible college', 82–4, 91

Jevons, F. R., 89, 90
Jewkes, J., 91

93

Index

journals, scientific, 22, 79
journalist, *see* science journalist

Keats, J., 19
Keble Martin, W., 45–7, 90
King, A., 89
Kuhn, T. S., 10, 89

Langrish, J., 90
learned societies, 18, 21–4, 79, 81
life sciences, 45–8
Los Angeles Times, 32
Lunar Society of Birmingham, 18

McElhenny, V. K., 90
Macleod, R. M., 89
Martin, D., 90
magazines, 39–41, 78–9; *see also* journals
Manchester Literary and Philosophical Society, 18
Mathias, P., 89
Meadows, D. H., 91
Meadows, D. L. 91
Meadowns, D. L., 91
mechanics institutes, 18
Medawar, Sir Peter (B.), 90, 91
medicine, 45
Metzger, N., 90
Michaelis, A. R., 91
Monod, J., 90
Morris, D., 47, 90

Nathe, L., 90
Natural Environment Research Council, 83
Nature, 39
Nature–Times news service, 8, 34
New Scientist, 17, 38–40
New York Times, 32
news, 6–9
newspapers, 29–38, 77–8; *see also* Press
Nunn, C., 25, 90

paradigm, 10
Pepper, J. H., 63, 91
Periodicals, 2; *see also* journals, magazines
Perlman, D., 30, 90
physical sciences, 48–50, 51
policy in newspapers, 30
pollution, 68, 69, 70
Popper, Sir Karl (R.), 44

presentation, newspaper, 35
Press, the, 29–38, 55
public schools, 18
Puritan ethic, 17
Pyke, M., 21, 90

radio, 2, 29, 56, 79; *see also* British Broadcasting Corporation
Ravetz, J. R., 91
Read, Sir Herbert, 89
Reid, R. W., 90
Ritchie-Calder, Baron (P. R. Ritchie-Calder), 90
Robinson, M., 90
Ronayne, J., 90
Rose, H., 91
Rose, S., 91
Royal Institute of Chemistry, 22
Royal Institution, 23–4, 80
Royal Society of London, 18, 81, 83

St Louis Post-Despatch, 32
Salomon, J.-J., 69, 91
San Francisco Chronicle, 30
science, cultural aspects, 2–3, 5, 6, 37, 86–7
problems of communicating, 14–17, 42–4, 63–6
public attitudes to, 19–20, 66–9
and society, 26–7, 51, 66–9, 70–1, 72–4, 76–7
science correspondent, 29–32; *see also* science journalist,
Science Journal, 39
science journalist and science journalism, 4, 9, 10, 11–13, 14–15, 25, 68–9, 74–5, 80–1, 84–5; *see also* science correspondent, science writer
science policy, 3
science writer, and science writing, 26, 27–8, 84–6; *see also* science correspondent, science journalist
Scientific American, 8
scientific papers, 44, 52
scientific reporting, quality of, 33 *et seq.*
scientists, communications between, 1
contribution to communication, 8–9, 79, 80
and journalist, 81–4

94

Index

Seaborg, G. T., 91
Shelley, P. B., 19
Smith, P. J., 91
Snow, Sir Charles (P.), 89
society publications, 40
Stillerman, D., 91

Tanders, J., 91
technology, 69– , 72
'technonature' and technostructure, 69
television, 2, 29, 52–62, 75, 79–80
  audience for science, 52, 56
  limitations, 60, 61
  and newspapers, 55–6
  producers, 56
  programme policy, 56, 57–8, 61–2
  strengths, 60–1
  viewers, 57
Tennyson, Sir Arthur, 19
The Times, 18, 34, 35, 75
Tinker, J., 90
Tomorrow's World, 59–60
trade and technical magazines, 40
Travers, M. W., 89
Tilling. L., 90

United States of America, 25
USSR, 20

Weinberg, S., 90
Wood, J., 90
Wordsworth, W., 19

Ziman, J., 89